Bioseparations
Principles and Techniques

Bioseparations
Principles and Techniques

Isaac Peterson
Maya Dixon

Kruger Brentt
Publishers
2026

Kruger Brentt Publishers UK. LTD.
Company Number 9728962

Regd. Office: 68 St Margarets Road, Edgware, Middlesex HA8 9UU

ISBN: 978-1-78715-603-6

For information on all our publications visit our website at http://krugerbrentt.com/

Preface

Bioseparations form the backbone of biotechnology and bioprocessing, encompassing the methods and technologies used to isolate, purify, and recover biological products from complex mixtures. From pharmaceuticals and enzymes to vaccines and biofuels, efficient bioseparation techniques are critical for ensuring product quality, safety, and economic viability. As biotechnology continues to expand across medicine, agriculture, and industrial applications, a thorough understanding of bioseparation principles and techniques is indispensable for researchers, students, and professionals in the field.

This book, Bioseparations: Principles and Techniques, provides a comprehensive guide to the fundamental concepts, methodologies, and practical applications of bioseparations. Designed for students, researchers, and practitioners of biotechnology, biochemistry, and related disciplines, it offers both theoretical insights and applied knowledge to address the challenges of isolating and purifying biomolecules.

The opening chapters introduce the essential principles of bioseparation, including the physicochemical properties of biomolecules, solubility, molecular size, charge, and affinity interactions, which determine the behavior of molecules during separation. It then explores a variety of separation techniques, including filtration, centrifugation, precipitation, chromatography, and membrane-based methods, explaining the mechanisms, advantages, and limitations of each approach.

Subsequent chapters focus on advanced techniques such as affinity chromatography, electrophoresis, ultrafiltration, and high-performance liquid chromatography (HPLC), highlighting their applications in industrial and research settings. Case studies demonstrate how bioseparation methods are employed in the production of biopharmaceuticals, enzymes, and other high-value biological products.

The book also addresses process optimization, scale-up considerations, and quality control aspects, emphasizing the integration of bioseparation strategies into larger bioprocess workflows. Additionally, emerging technologies such as continuous processing, automation, and microfluidic systems are discussed, reflecting the cutting-edge developments that are shaping the future of bioseparations.

Bioseparations: Principles and Techniques serves as an essential resource for anyone seeking to master the science and technology behind the purification of biological products. By combining clear explanations of fundamental concepts with practical guidance and real-world applications, this book equips readers with the tools necessary to design, implement, and optimize effective bioseparation processes.

Isaac Peterson

Maya Dixon

Contents

CHAPTER 1

An Overview of Bioseparations

Bioseparation is an essential and evolving field that plays a critical role in the extraction and purification of biological materials, such as proteins, enzymes, nucleic acids, and other biomolecules. The applications of bioseparations span a wide range of industries, including pharmaceuticals, biotechnology, food and beverage, environmental management, and agricultural sectors. The goal of bioseparation processes is to separate and purify target biomolecules from complex mixtures while maintaining their biological activity and integrity. In recent years, bioseparation technologies have become increasingly important due to the growing demand for high-purity bio-based products, as well as the need to meet regulatory standards and environmental sustainability criteria.

1.1 PRINCIPLES OF BIOSEPARATION

At its core, bioseparation is based on the principles of selective separation, where the biomolecule of interest is isolated from a mixture by exploiting differences in their physical, chemical, or biological properties. The primary goal is to achieve a high degree of purity while minimizing the loss of activity or yield of the target compound. Commonly utilized separation methods include filtration, centrifugation, chromatography, membrane processes, and precipitation. Each technique capitalizes on different principles, such as size, charge, hydrophobicity, or affinity, to separate biomolecules from other components in a mixture.

1.2 TYPES OF BIOSEPARATION METHODS

There are several bioseparation techniques, each with its strengths, limitations, and specific applications. Filtration is one of the most basic methods and is widely used in the first stages of separation. It relies on physical barriers such as membranes to separate particles of different sizes. Ultrafiltration and microfiltration are commonly used in the biotechnology industry to remove small molecules or larger particles, such as cells or cell debris, from a solution.

Centrifugation is another widely used technique, which separates biomolecules based on their size, shape, and density. By applying centrifugal force, it enables the separation of components in a mixture into distinct layers. This method is especially useful for separating cells, cellular components, and biomolecules that differ in their sedimentation properties.

Chromatography is one of the most powerful and versatile bioseparation techniques. It is based on the differential distribution of components between a stationary phase and a mobile phase. There are several types of chromatography, such as affinity chromatography, ion-exchange chromatography, and size-exclusion chromatography, each targeting specific properties of the biomolecule, such as its affinity for certain ligands, charge, or size. This method is widely employed for purifying proteins, enzymes, and other biopharmaceuticals.

Membrane-based separation techniques, such as reverse osmosis, nanofiltration, and electrodialysis, are also gaining traction in bioseparations. These methods involve the use of semi-permeable membranes to selectively allow certain ions or molecules to pass through while rejecting others. They are particularly useful in processes such as desalting, concentration, and purification of biomolecules.

1.3 APPLICATIONS OF BIOSEPARATION

Bioseparations are critical in a range of applications, particularly in the production of pharmaceuticals and biotechnology products. In the pharmaceutical industry, the need for high-purity biopharmaceuticals, such as monoclonal antibodies, recombinant proteins, and vaccines, has driven the development of advanced bioseparation techniques. The production of these biopharmaceuticals often involves the use of genetically modified organisms, which produce the desired product in large quantities. However, these products need to be isolated from the culture media and other by-products, requiring effective and scalable bioseparation methods.

In the field of biotechnology, bioseparation plays a crucial role in the production of enzymes, biofuels, and other biologically derived chemicals. The enzymes used in industrial processes often require high purity to maintain their effectiveness and stability. The purification of enzymes and other biomolecules from microbial cultures, plant, or animal sources is an essential part of biotechnological production.

In food and beverage industries, bioseparation methods are used to extract valuable ingredients such as proteins, flavor compounds, and essential oils from natural sources. For example, the separation of protein from soybeans, whey, or egg whites is commonly performed to produce food ingredients with specific functional properties.

Environmental applications also benefit from bioseparations, particularly in water treatment and waste management. Bioseparation technologies are used to purify water by removing contaminants such as heavy metals, organic pollutants, and bacteria. In waste treatment, bioseparation processes can help recover valuable biomolecules from industrial effluents or waste streams, thus reducing environmental impact.

1.4 CHALLENGES IN BIOSEPARATION

Despite the advances in bioseparation technologies, several challenges remain. One of the primary challenges is the complexity and heterogeneity of biological mixtures. Unlike traditional chemical separations, biological mixtures often contain a wide range of components with similar sizes, charges, or properties, making the separation process more challenging. In addition, maintaining the stability and activity of biomolecules during the separation process is often a concern. Many biomolecules are sensitive to changes in temperature, pH, and shear forces, and any damage to their structure can result in the loss of their activity.

Scalability is another challenge in bioseparation. While laboratory-scale separations can achieve high purity and yield, scaling up these processes to industrial levels while maintaining efficiency and cost-effectiveness is often a complex task. In large-scale operations, factors such as equipment size, throughput, and cost of consumables can significantly impact the feasibility of a bioseparation process.

Moreover, the integration of different bioseparation technologies to create a multi-step process often presents challenges in terms of process optimization and cost reduction. The combination of several techniques, such as centrifugation followed by chromatography or filtration, can lead to increased process time, operational complexity, and higher production costs.

1.5 FUTURE DIRECTIONS IN BIOSEPARATION

The future of bioseparations holds great promise, with ongoing research focused on improving efficiency, reducing costs, and enhancing the scalability of bioseparation processes. The development of new materials, such as advanced membranes with enhanced selectivity and stability, could help address many of the challenges in bioseparation. For example, the development of high-capacity affinity resins or novel chromatography media may significantly improve the speed and effectiveness of purification processes.

Another promising direction in bioseparation is the use of microfluidics, which involves the manipulation of fluids at the microscale. Microfluidic devices allow for the precise control of fluid flow and separation processes, enabling faster and more efficient bioseparations. This technology could be particularly useful for small-scale or lab-on-a-chip applications, where the rapid separation of biomolecules is required.

Additionally, advances in computational modeling and artificial intelligence may allow for the better prediction and optimization of bioseparation processes. By modeling the interactions between biomolecules and separation media, researchers can design more efficient processes that minimize waste and improve product yield.

1.6 CONCLUSION

Bioseparation is a vital and dynamic field that enables the purification and isolation of biomolecules for a wide range of applications in biotechnology, pharmaceuticals, food industries, and environmental management. While there are significant challenges to overcome, including complexity, scalability, and cost, advances in technology and research are poised to revolutionize the field. With the increasing demand for high-quality, bio-based products, bioseparation technologies will continue to evolve and play a pivotal role in meeting these needs.

CHAPTER 2

Cell Disruption

2.1 INTRODUCTION TO CELL DISRUPTION

Cell disruption, sometimes referred to as digestion, is a method or process for releasing biological molecules from inside a cell.

2.1.1 Methods

The production of biologically interesting molecules using cloning and culturing methods allows the study and manufacture of relevant molecules. Except for excreted molecules, cells producing molecules of interest must be disrupted. This page discusses various methods. Another method of disruption is called cell unroofing.

Bead Method

A common laboratory-scale mechanical method for cell disruption uses glass, ceramic, or steel beads, 0.1–2 mm (0.004–0.08 in) in diameter, mixed with a sample suspended in an aqueous solution. First developed by Tim Hopkins in the late 1970s, the sample and bead mix is subjected to high level agitation by stirring or shaking. Beads collide with the cellular sample, cracking open the cell to release the intracellular components. Unlike some other methods, mechanical shear is moderate during homogenization resulting in excellent membrane or subcellular preparations. The method, often called "bead beating", works well for all types of cellular material - from spores to animal and plant tissues. It is the most widely used method of yeast lysis, and can yield breakage of well over 50% (up to 95%). It has the advantage over other mechanical cell disruption methods of being able to disrupt very small sample sizes, process many samples at a time with no cross-contamination concerns, and does not release potentially harmful aerosols in the process.

In the simplest example of the method, an equal volume of beads are added to a cell or tissue suspension in a test tube and the sample is vigorously mixed on a common laboratory vortex mixer. While processing times are slow, taking 3–10 times longer than that in specialty shaking machines, it works well for easily disrupted cells and is inexpensive. Successful bead beating is dependent not only on design features of the shaking machine (which take into consideration shaking oscillations frequency, shaking throw or distance, shaking orientation and vial orientation), but also the selection of correct bead size (0.1–6 mm (0.004–0.2 in) diameter), bead composition (glass, ceramic, steel) and bead load in the vial.

In most laboratories, bead beating is done in batch sizes of one to twenty-four sealed, plastic vials or centrifuge tubes. The sample and tiny beads are agitated at about 2000 oscillations per minute in specially designed reciprocating shakers driven by high power electric motors. Cell disruption is complete in 1–3 minutes of shaking. Significantly faster rates of cell disruption are achieved with a bead beater variation called SoniBeast. Differing from conventional machines, it agitates the beads using a vortex motion at 20,000 oscillations per minute. Larger bead beater machines that hold deep-well microtiter plates also shorten process times, as do Bead Dispensers designed to quickly load beads into multiple vials or microplates. Pre-loaded vials and microplates are also available.

All high energy bead beating machines warm the sample about 10 degrees per minute. This is due to frictional collisions of the beads during homogenization. Cooling of the sample during or after bead beating may be necessary to prevent damage to heat-sensitive proteins such as enzymes. Sample warming can be controlled by bead beating for short time intervals with cooling on ice between each interval, by processing vials in pre-chilled aluminum vial holders or by circulating gaseous coolant through the machine during bead beating.

A different bead beater configuration, suitable for larger sample volumes, uses a rotating fluorocarbon rotor inside a 15, 50 or 200 ml chamber to agitate the beads. In this configuration, the chamber can be surrounded by a static cooling jacket. Using this same rotor/chamber configuration, large commercial machines are available to process many liters of cell suspension. Currently, these machines are limited to processing unicellular organisms such as yeast, algae and bacteria.

Cryopulverization

Samples with a tough extracellular matrix, such as animal connective tissue, some tumor biopsy samples, venous tissue, cartilage, seeds, etc., are reduced to a fine powder by impact pulverization at liquid nitrogen temperatures. This technique, known as cryopulverization, is based on the fact that biological samples containing a significant fraction of water become brittle at extremely cold temperatures. This technique was first described by Smucker and Pfister in 1975, who referred to the technique as cryo-impacting. The authors demonstrated cells are effectively broken by this method, confirming by phase and electron microscopy that breakage planes cross cell walls and cytoplasmic membranes.

The technique can be done by using a mortar and pestle cooled to liquid nitrogen temperatures, but use of this classic apparatus is laborious and sample loss is often a concern. Specialised stainless steel pulverizers generically known as Tissue Pulverizers are also available for this purpose. They require less manual effort, give good sample recovery and are easy to clean between samples. Advantages of this technique are higher yields of proteins and nucleic acids from small, hard tissue samples - especially when used as a preliminary step to mechanical or chemical/ solvent cell disruption methods.

High Pressure Cell Disruption

Since the 1940s high pressure has been used as a method of cell disruption, most notably by the French Pressure Cell Press, or French Press for short. This method was developed by Charles Stacy French and utilises high pressure to force cells through a narrow orifice, causing the cells to lyse due to the shear forces experienced across the pressure differential. While French Presses have become a staple item in many microbiology laboratories, their production has been largely discontinued, leading to a resurgence in alternate applications of similar technology.

Modern physical cell disruptors typically operate via either pneumatic or hydraulic pressure. Although pneumatic machines are typically lower cost, their performance can be unreliable due to variations in the processing pressure throughout the stroke of the air pump. It is generally considered that hydraulic machines offer superior lysing ability, especially when processing harder to break samples such as yeast or Gram-positive bacteria, due to their ability to maintain constant pressure throughout the piston stroke. As the French Press, which is operated by hydraulic pressure, is capable of over 90% lysis of most commonly used cell types it is often taken as the gold standard in lysis performance and modern machines are often compared against it not only in terms of lysis efficiency but also in terms of safety and ease of use. Some manufacturers are also trying to improve on the traditional design by altering properties within these machines other than the pressure driving the sample through the orifice. One such example is Constant Systems, who have recently shown that their Cell Disruptors not only match the performance of a traditional French Press, but also that they are striving towards attaining the same results at a much lower power.

Pressure Cycling Technology ("PCT"). PCT is a patented, enabling technology platform that uses alternating cycles of hydrostatic pressure between ambient and ultra-high levels (up to 90,000 psi) to safely, conveniently and reproducibly control the actions of molecules in biological samples, e.g., the rupture (lysis) of cells and tissues from human, animal, plant, and microbial sources, and the inactivation of pathogens. PCT-enhanced systems (instruments and consumables) address some challenging problems inherent in biological sample preparation. PCT advantages include: (a) extraction and recovery of more membrane proteins, (b) enhanced protein digestion, (c) differential lysis in a mixed sample base, (d) pathogen inactivation, (e) increased DNA detection, and (f) exquisite sample preparation process control.

The Microfluidizer method used for cell disruption strongly influences the physicochemical properties of the lysed cell suspension, such as particle size, viscosity, protein yield and enzyme activity. In recent years the Microfluidizer method has gained popularity in cell disruption due to its ease of use and efficiency at disrupting many different kinds of cells. The Microfluidizer technology was licensed from a company called Arthur D. Little and was first developed and utilized in the 1980s, initially starting as a tool for liposome creation. It has since been used in other applications such as cell disruption nanoemulsions, and solid particle size reduction, among others.

By using microchannels with fixed geometry, and an intensifier pump, high shear rates are generated that rupture the cells. This method of cell lysis can yield breakage of over 90% of E. coli cells.

Many proteins are extremely temperature-sensitive, and in many cases can start to denature at temperatures of only 4 degrees Celsius. Within the microchannels, temperatures exceed 4 degrees Celsius, but the machine is designed to cool quickly so that the time the cells are exposed to elevated temperatures is extremely short (residence time 25 ms-40 ms). Because of this effective temperature control, the Microfluidizer yields higher levels of active proteins and enzymes than other mechanical methods when the proteins are temperature-sensitive.

Viscosity changes are also often observed when disrupting cells. If the cell suspension viscosity is high, it can make downstream handling—such as filtration and accurate pipetting—quite difficult. The viscosity changes observed with a Microfluidizer are relatively low, and decreases with further additional passes through the machine.

In contrast to other mechanical disruption methods the Microfluidizer breaks the cell membranes efficiently but gently, resulting in relatively large cell wall fragments (450 nm), and thus making it easier to separate the cell contents. This can lead to shorter filtration times and better centrifugation separation.

Microfluidizer technology scales from one milliliter to thousands of liters.

Nitrogen Decompression

For nitrogen decompression, large quantities of nitrogen are first dissolved in the cell under high pressure within a suitable pressure vessel. Then, when the gas pressure is suddenly released, the nitrogen comes out of the solution as expanding bubbles that stretch the membranes of each cell until they rupture and release the contents of the cell.

Nitrogen decompression is more protective of enzymes and organelles than ultrasonic and mechanical homogenizing methods and compares favorably to the controlled disruptive action obtained in a PTFE and glass mortar and pestle homogenizer. While other disruptive methods depend upon friction or a mechanical shearing action that generate heat, the nitrogen decompression procedure is accompanied by an adiabatic expansion that cools the sample instead of heating it.

The blanket of inert nitrogen gas that saturates the cell suspension and the homogenate offers protection against oxidation of cell components. Although other gases: carbon dioxide, nitrous oxide, carbon monoxide and compressed air have been used in this technique, nitrogen is preferred because of its non-reactive nature and because it does not alter the pH of the suspending medium. In addition, nitrogen is preferred because it is generally available at low cost and at pressures suitable for this procedure.

Once released, subcellular substances are not exposed to continued attrition that might denature the sample or produce unwanted damage. There is no need to watch for a peak between enzyme activity and percent disruption. Since nitrogen bubbles are generated within each cell, the same disruptive force is applied uniformly throughout the sample, thus ensuring unusual uniformity in the product. Cell-free homogenates can be produced.

The technique is used to homogenize cells and tissues, release intact organelles, prepare cell membranes, release labile biochemicals, and produce uniform and repeatable homogenates without subjecting the sample to extreme chemical or physical stress.

The method is particularly well suited for treating mammalian and other membrane-bound cells. It has also been used successfully for treating plant cells, for releasing virus from fertilized eggs and for treating fragile bacteria. It is not recommended for untreated bacterial cells. Yeast, fungus, spores and other materials with tough cell walls do not respond well to this method.

2.2 UNDERSTANDING CELL MEMBRANE STRUCTURE

The cell membrane (also known as the plasma membrane or cytoplasmic membrane, and historically referred to as the plasmalemma) is a biological membrane that separates and protects the interior of a cell from the outside environment (the extracellular space). The cell membrane consists of a lipid bilayer, made up of two layers of phospholipids with cholesterols (a lipid component) interspersed between them, maintaining appropriate membrane fluidity at various temperatures. The membrane also contains membrane proteins, including integral proteins that span the membrane and serve as membrane transporters, and peripheral proteins that loosely attach to the outer (peripheral) side of the cell membrane, acting as enzymes to facilitate interaction with the cell's environment. Glycolipids embedded in the outer lipid layer serve a similar purpose.

The cell membrane controls the movement of substances in and out of a cell, being selectively permeable to ions and organic molecules. In addition, cell membranes are involved in a variety of cellular processes such as cell adhesion, ion conductivity, and cell signalling and serve as the attachment surface for several extracellular structures, including the cell wall and the carbohydrate layer called the glycocalyx, as well as the intracellular network of protein fibers called the cytoskeleton. In the field of synthetic biology, cell membranes can be artificially reassembled.

2.2.1 History

Robert Hooke's discovery of cells in 1665 led to the proposal of the cell theory. Initially it was believed that all cells contained a hard cell wall since only plant cells could be observed at the time. Microscopists focused on the cell wall for well over 150 years until advances in microscopy were made. In the early 19th century, cells were recognized as being separate entities, unconnected, and bound by individual cell walls after it was found that plant cells could be separated. This theory extended to include animal cells to suggest a universal mechanism for cell protection and development.

By the second half of the 19th century, microscopy was still not advanced enough to make a distinction between cell membranes and cell walls. However, some microscopists correctly identified at this time that while invisible, it could be inferred that cell membranes existed in animal cells due to intracellular movement of components internally but not externally and that membranes were not the equivalent of a plant cell wall. It was also inferred that cell membranes were not vital components to all cells. Many refuted the existence of a cell membrane still towards the end of the 19th century. In 1890, a revision to the cell theory stated that cell membranes existed, but were merely secondary structures. It was not until later studies with osmosis and permeability that cell membranes gained more recognition. In 1895, Ernest Overton proposed that cell membranes were made of lipids.

The lipid bilayer hypothesis, proposed in 1925 by Gorter and Grendel, created speculation in the description of the cell membrane bilayer structure based on crystallographic studies and soap bubble observations. In an attempt to accept or reject the hypothesis, researchers measured membrane thickness. These researchers extracted the lipid from human red blood cells and measured the amount of surface area the lipid would cover when spread over the surface of the water. Since mature mammalian red blood cells lack both nuclei and cytoplasmic organelles, the plasma membrane is the only lipid-containing structure in the cell.

Consequently, all of the lipids extracted from the cells can be assumed to have resided in the cells' plasma membranes. The ratio of the surface area of water covered by the extracted lipid to the surface area calculated for the red blood cells from which the lipid was 2:1(approx) and they concluded that the plasma membrane contains a lipid bilayer.

In 1925 it was determined by Fricke that the thickness of erythrocyte and yeast cell membranes ranged between 3.3 and 4 nm, a thickness compatible with a lipid monolayer. The choice of the dielectric constant used in these studies was called into question but future tests could not disprove the results of the initial experiment. Independently, the leptoscope was invented in order to measure very thin membranes by comparing the intensity of light reflected from a sample to the intensity of a membrane standard of known thickness. The instrument could resolve thicknesses that depended on pH measurements and the presence of membrane proteins that ranged from 8.6 to 23.2 nm, with the lower measurements supporting the lipid bilayer hypothesis. Later in the 1930s, the membrane structure model developed in general agreement to be the paucimolecular model of Davson and Danielli (1935). This model was based on studies of surface tension between oils and echinoderm eggs. Since the surface tension values appeared to be much lower than would be expected for an oil-water interface, it was assumed that some substance was responsible for lowering the interfacial tensions in the surface of cells. It was suggested that a lipid bilayer was in between two thin protein layers. The paucimolecular model immediately became popular and it dominated cell membrane studies for the following 30 years, until it became rivaled by the fluid mosaic model of Singer and Nicolson (1972).

Despite the numerous models of the cell membrane proposed prior to the fluid mosaic model, it remains the primary archetype for the cell membrane long after its inception in the 1970s. Although the fluid mosaic model has been modernized to detail contemporary discoveries, the basics have remained constant: the membrane is a lipid bilayer composed of hydrophilic exterior heads and a hydrophobic interior where proteins can interact with hydrophilic heads through polar interactions, but proteins that span the bilayer fully or partially have hydrophobic amino acids that interact with the non-polar lipid interior. The fluid mosaic model not only provided an accurate representation of membrane mechanics, it enhanced the study of hydrophobic forces, which would later develop into an essential descriptive limitation to describe biological macromolecules.

For many centuries, the scientists cited disagreed with the significance of the structure they were seeing as the cell membrane. For almost two centuries, the membranes were seen but mostly disregarded as an important structure with cellular function. It was not until the 20th century that the significance of the cell membrane as it was acknowledged. Finally, two scientists Gorter and Grendel (1925) made the discovery that the membrane is "lipid-based". From this, they furthered the idea that this structure would have to be in a formation that mimicked layers. Once studied further,

it was found by comparing the sum of the cell surfaces and the surfaces of the lipids, a 2:1 ratio was estimated; thus, providing the first basis of the bilayer structure known today. This discovery initiated many new studies that arose globally within various fields of scientific studies, confirming that the structure and functions of the cell membrane are widely accepted.

The structure has been variously referred to by different writers as the ectoplast (de Vries, 1885), Plasmahaut (plasma skin, Pfeffer, 1877, 1891), Hautschicht (skin layer, Pfeffer, 1886; used with a different meaning by Hofmeister, 1867), plasmatic membrane (Pfeffer, 1900), plasma membrane, cytoplasmic membrane, cell envelope and cell membrane. Some authors who did not believe that there was a functional permeable boundary at the surface of the cell preferred to use the term plasmalemma (coined by Mast, 1924) for the external region of the cell.

2.2.2 Composition

Cell membranes contain a variety of biological molecules, notably lipids and proteins. Composition is not set, but constantly changing for fluidity and changes in the environment, even fluctuating during different stages of cell development. Specifically, the amount of cholesterol in human primary neuron cell membrane changes, and this change in composition affects fluidity throughout development stages.

Material is incorporated into the membrane, or deleted from it, by a variety of mechanisms:

- Fusion of intracellular vesicles with the membrane (exocytosis) not only excretes the contents of the vesicle but also incorporates the vesicle membrane's components into the cell membrane. The membrane may form blebs around extracellular material that pinch off to become vesicles (endocytosis).
- If a membrane is continuous with a tubular structure made of membrane material, then material from the tube can be drawn into the membrane continuously.
- Although the concentration of membrane components in the aqueous phase is low (stable membrane components have low solubility in water), there is an exchange of molecules between the lipid and aqueous phases.

Lipids

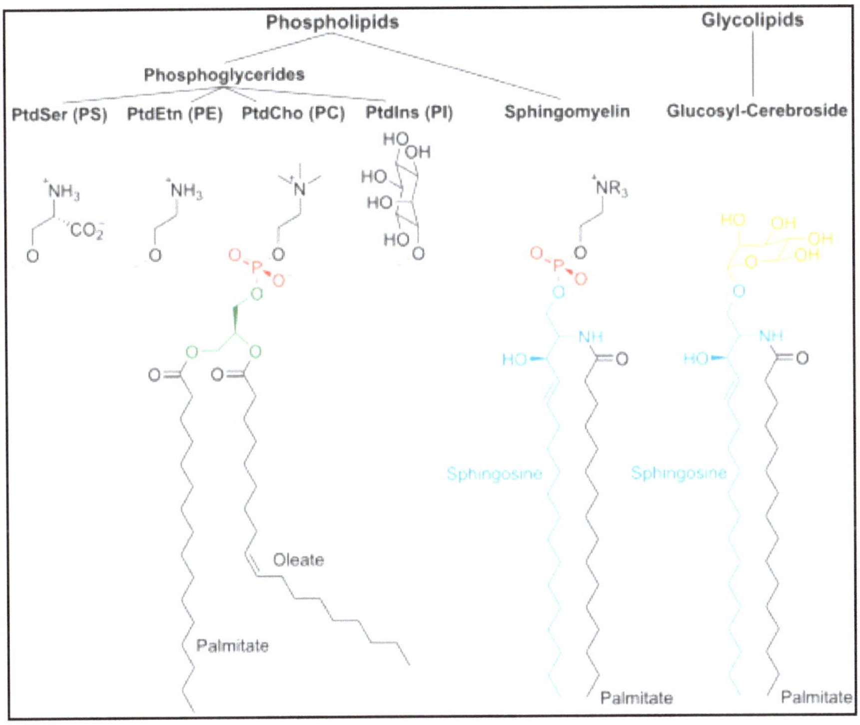

Fig: Examples of the Major Membrane Phospholipids and Glycolipids: Hosphatidylcholine (PtdCho), Phosphatidylethanolamine (PtdEtn), Phosphatidylinositol (PtdIns), Phosphatidylserine (PtdSer)

The cell membrane consists of three classes of amphipathic lipids: phospholipids, glycolipids, and sterols. The amount of each depends upon the type of cell, but in the majority of cases phospholipids are the most abundant, often contributing for over 50% of all lipids in plasma membranes. Glycolipids only account for a minute amount of about 2% and sterols make up the rest. In red blood cell studies, 30% of the plasma membrane is lipid. However, for the majority of eukaryotic cells, the composition of plasma membranes is about half lipids and half proteins by weight.

The fatty chains in phospholipids and glycolipids usually contain an even number of carbon atoms, typically between 16 and 20. The 16- and 18-carbon fatty acids are the most common. Fatty acids may be saturated or unsaturated, with the configuration of the double bonds nearly always “cis”. The length and the degree of unsaturation of fatty acid chains have a profound effect on membrane fluidity as unsaturated lipids create a kink, preventing the fatty acids from packing together as tightly, thus decreasing the melting temperature (increasing the fluidity) of the membrane. The ability of some organisms to regulate the fluidity of their cell membranes by altering lipid composition is called homeoviscous adaptation.

The entire membrane is held together via non-covalent interaction of hydrophobic tails, however the structure is quite fluid and not fixed rigidly in place. Under physiological conditions phospholipid molecules in the cell membrane are in the liquid crystalline state. It means the lipid molecules are free to diffuse and exhibit rapid lateral diffusion along the layer in which they are present. However, the exchange of phospholipid molecules between intracellular and extracellular leaflets of the bilayer is a very slow process. Lipid rafts and caveolae are examples of cholesterol-enriched microdomains in the cell membrane. Also, a fraction of the lipid in direct contact with integral membrane proteins, which is tightly bound to the protein surface is called annular lipid shell; it behaves as a part of protein complex.

Cholesterol is normally found dispersed in varying degrees throughout cell membranes, in the irregular spaces between the hydrophobic tails of the membrane lipids, where it confers a stiffening and strengthening effect on the membrane. Additionally, the amount of cholesterol in biological membranes varies between organisms, cell types, and even in individual cells. Cholesterol, a major component of plasma membranes, regulates the fluidity of the overall membrane, meaning that cholesterol controls the amount of movement of the various cell membrane components based on its concentrations. In high temperatures, cholesterol inhibits the movement of phospholipid fatty acid chains, causing a reduced permeability to small molecules and reduced membrane fluidity. The opposite is true for the role of cholesterol in cooler temperatures. Cholesterol production, and thus concentration, is up-regulated (increased) in response to cold temperature. At cold temperatures, cholesterol interferes with fatty acid chain interactions. Acting as antifreeze, cholesterol maintains the fluidity of the membrane. Cholesterol is more abundant in cold-weather animals than warm-weather animals. In plants, which lack cholesterol, related compounds called sterols perform the same function as cholesterol.

Phospholipids Forming Lipid Vesicles

Lipid vesicles or liposomes are approximately spherical pockets that are enclosed by a lipid bilayer. These structures are used in laboratories to study the effects of chemicals in cells by delivering these chemicals directly to the cell, as well as getting more insight into cell membrane permeability. Lipid vesicles and liposomes are formed by first suspending a lipid in an aqueous solution then agitating the mixture through sonication, resulting in a vesicle. Measuring the rate of efflux from the inside of the vesicle to the ambient solution allows researchers to better understand membrane permeability. Vesicles can be formed with molecules and ions inside the vesicle by forming the vesicle with the desired molecule or ion present in the solution. Proteins can also be embedded into the membrane through solubilizing the desired proteins in the presence of detergents and attaching them to the phospholipids in which the liposome is formed. These provide researchers with a tool to examine various membrane protein functions.

Carbohydrates

Plasma membranes also contain carbohydrates, predominantly glycoproteins, but with some glycolipids (cerebrosides and gangliosides). Carbohydrates are important in the role of cell-cell recognition in eukaryotes; they are located on the surface of the cell where they recognize host cells and share information. Viruses that bind to cells using these receptors cause an infection.

For the most part, no glycosylation occurs on membranes within the cell; rather generally glycosylation occurs on the extracellular surface of the plasma membrane. The glycocalyx is an important feature in all cells, especially epithelia with microvilli. Recent data suggest the glycocalyx participates in cell adhesion, lymphocyte homing, and many others. The penultimate sugar is galactose and the terminal sugar is sialic acid, as the sugar backbone is modified in the Golgi apparatus. Sialic acid carries a negative charge, providing an external barrier to charged particles.

Proteins

Type	Description	Examples
Integral proteins or transmembrane proteins	Span the membrane and have a hydrophilic cytosolic domain, which interacts with internal molecules, a hydrophobic membrane-spanning domain that anchors it within the cell membrane, and a hydrophilic extracellular domain that interacts with external molecules. The hydrophobic domain consists of one, multiple, or a combination of α-helices and β sheet protein motifs.	Ion channels, proton pumps, G protein-coupled receptor
Lipid anchored proteins	Covalently bound to single or multiple lipid molecules; hydrophobically insert into the cell membrane and anchor the protein. The protein itself is not in contact with the membrane.	G proteins

Peripheral proteins	Attached to integral membrane proteins, or associated with peripheral regions of the lipid bilayer. These proteins tend to have only temporary interactions with biological membranes, and once reacted, the molecule dissociates to carry on its work in the cytoplasm.	Some enzymes, some hormones

The cell membrane has large content of proteins, typically around 50% of membrane volume These proteins are important for the cell because they are responsible for various biological activities. Approximately a third of the genes in yeast code specifically for them, and this number is even higher in multicellular organisms. Membrane proteins consist of three main types: integral proteins, peripheral proteins, and lipid-anchored proteins.

As shown in the adjacent table, integral proteins are amphipathic transmembrane proteins. Examples of integral proteins include ion channels, proton pumps, and g-protein coupled receptors. Ion channels allow inorganic ions such as sodium, potassium, calcium, or chlorine to diffuse down their electrochemical gradient across the lipid bilayer through hydrophilic pores across the membrane. The electrical behavior of cells (i.e. nerve cells) is controlled by ion channels. Proton pumps are protein pumps that are embedded in the lipid bilayer that allow protons to travel through the membrane by transferring from one amino acid side chain to another. Processes such as electron transport and generating ATP use proton pumps. A G-protein coupled receptor is a single polypeptide chain that crosses the lipid bilayer seven times responding to signal molecules (i.e. hormones and neurotransmitters). G-protein coupled receptors are used in processes such as cell to cell signaling, the regulation of the production of cAMP, and the regulation of ion channels.

The cell membrane, being exposed to the outside environment, is an important site of cell–cell communication. As such, a large variety of protein receptors and identification proteins, such as antigens, are present on the surface of the membrane. Functions of membrane proteins can also include cell–cell contact, surface recognition, cytoskeleton contact, signaling, enzymatic activity, or transporting substances across the membrane.

Most membrane proteins must be inserted in some way into the membrane. For this to occur, an N-terminus "signal sequence" of amino acids directs proteins to the endoplasmic reticulum, which inserts the proteins into a lipid bilayer. Once inserted, the proteins are then transported to their final destination in vesicles, where the vesicle fuses with the target membrane.

2.2.3 Function

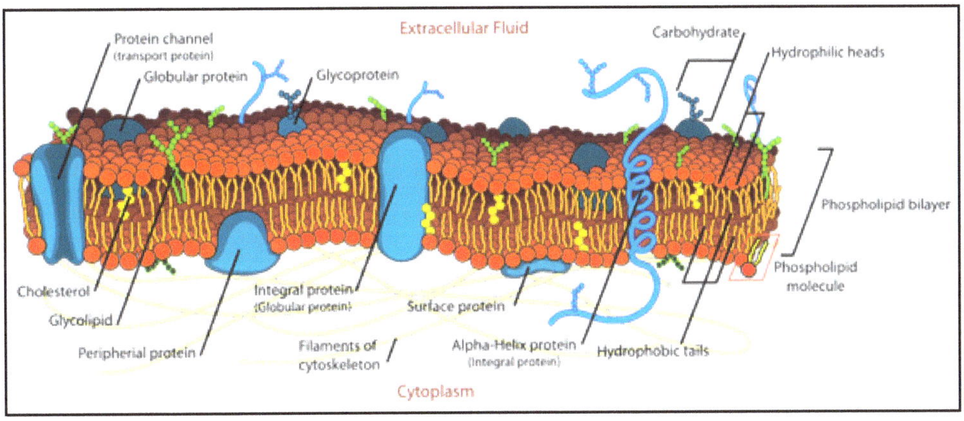

Fig: A Detailed Diagram of the Cell Membrane

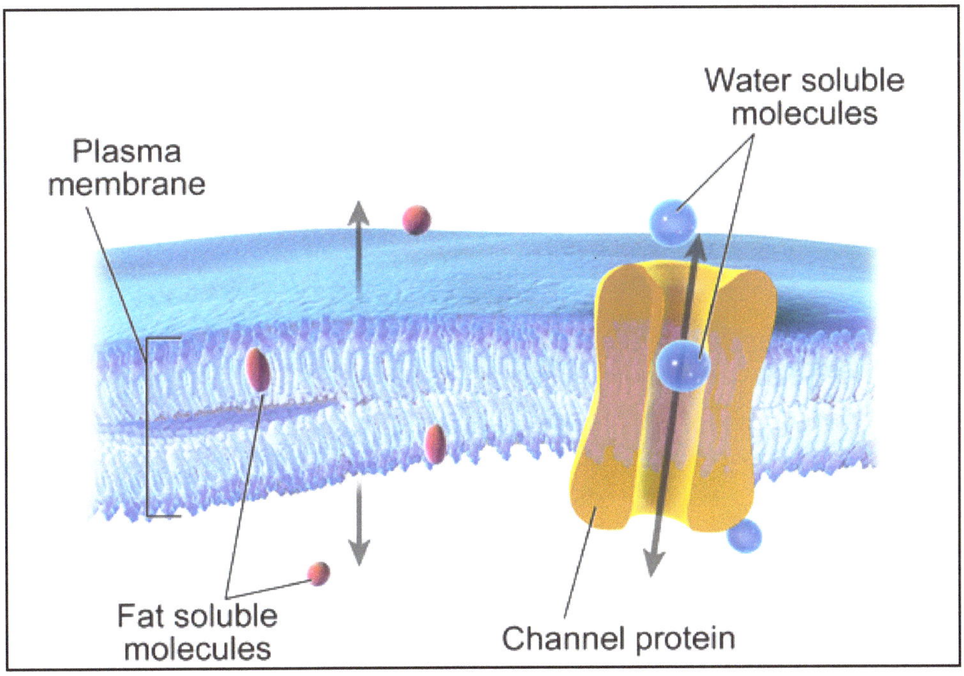

Fig: Illustration Depicting Cellular Diffusion

The cell membrane surrounds the cytoplasm of living cells, physically separating the intracellular components from the extracellular environment. The cell membrane also plays a role in anchoring the cytoskeleton to provide shape to the cell, and in attaching to the extracellular matrix and other cells to hold them together to form tissues. Fungi, bacteria, most archaea, and plants also have a cell wall, which provides a mechanical support to the cell and precludes the passage of larger molecules.

The cell membrane is selectively permeable and able to regulate what enters and exits the cell, thus facilitating the transport of materials needed for survival. The movement of substances across the membrane can be achieved by either passive transport, occurring without the input of cellular energy, or by active transport, requiring the cell to expend energy in transporting it. The membrane also maintains the cell potential. The cell membrane thus works as a selective filter that allows only certain things to come inside or go outside the cell. The cell employs a number of transport mechanisms that involve biological membranes:

- **Passive Osmosis and Diffusion:** Some substances (small molecules, ions) such as carbon dioxide (CO_2) and oxygen (O_2), can move across the plasma membrane by diffusion, which is a passive transport process. Because the membrane acts as a barrier for certain molecules and ions, they can occur in different concentrations on the two sides of the membrane. Diffusion occurs when small molecules and ions move freely from high concentration to low concentration in order to equilibrate the membrane. It is considered a passive transport process because it does not require energy and is propelled by the concentration gradient created by each side of the membrane. Such a concentration gradient across a semipermeable membrane sets up an osmotic flow for the water. Osmosis, in biological systems involves a solvent, moving through a semipermeable membrane similarly to passive diffusion as the solvent still moves with the concentration gradient and requires no energy. While water is the most common solvent in cell, it can also be other liquids as well as supercritical liquids and gases.
- **Transmembrane Protein Channels and Transporters:** Transmembrane proteins extend through the lipid bilayer of the membranes; they function on both sides of the membrane to transport molecules across it. Nutrients, such as sugars or amino acids, must enter the cell, and certain products of metabolism must leave the cell. Such molecules can diffuse passively through protein channels such as aquaporins in facilitated diffusion or are pumped across the membrane by transmembrane transporters. Protein channel proteins, also called permeases, are usually quite specific, and they only recognize and transport a limited variety of chemical substances, often limited to a single substance. Another example of a transmembrane protein is a cell-surface receptor, which allow cell signaling molecules to communicate between cells.
- **Endocytosis:** Endocytosis is the process in which cells absorb molecules by engulfing them. The plasma membrane creates a small deformation

inward, called an invagination, in which the substance to be transported is captured. This invagination is caused by proteins on the outside on the cell membrane, acting as receptors and clustering into depressions that eventually promote accumulation of more proteins and lipids on the cytosolic side of the membrane. The deformation then pinches off from the membrane on the inside of the cell, creating a vesicle containing the captured substance. Endocytosis is a pathway for internalizing solid particles ("cell eating" or phagocytosis), small molecules and ions ("cell drinking" or pinocytosis), and macromolecules. Endocytosis requires energy and is thus a form of active transport.

- **Exocytosis:** Just as material can be brought into the cell by invagination and formation of a vesicle, the membrane of a vesicle can be fused with the plasma membrane, extruding its contents to the surrounding medium. This is the process of exocytosis. Exocytosis occurs in various cells to remove undigested residues of substances brought in by endocytosis, to secrete substances such as hormones and enzymes, and to transport a substance completely across a cellular barrier. In the process of exocytosis, the undigested waste-containing food vacuole or the secretory vesicle budded from Golgi apparatus, is first moved by cytoskeleton from the interior of the cell to the surface. The vesicle membrane comes in contact with the plasma membrane. The lipid molecules of the two bilayers rearrange themselves and the two membranes are, thus, fused. A passage is formed in the fused membrane and the vesicles discharges its contents outside the cell.

2.2.4 Prokaryotes

Prokaryotes are divided into two different groups, Archaea and Bacteria, with bacteria dividing further into gram-positive and gram-negative. Gram-negative bacteria have both a plasma membrane and an outer membrane separated by periplasm; however, other prokaryotes have only a plasma membrane. These two membranes differ in many aspects. The outer membrane of the gram-negative bacteria differs from other prokaryotes due to phospholipids forming the exterior of the bilayer, and lipoproteins and phospholipids forming the interior. The outer membrane typically has a porous quality due to its presence of membrane proteins, such as gram-negative porins, which are pore-forming proteins. The inner plasma membrane is also generally symmetric whereas the outer membrane is asymmetric because of proteins such as the aforementioned.

Also, for the prokaryotic membranes, there are multiple things that can affect the fluidity. One of the major factors that can affect the fluidity is fatty acid composition. For example, when the bacteria Staphylococcus aureus was grown at 37 °C for 24 h, the membrane exhibited a more fluid state instead of a gel-like state. This supports the concept that in higher temperatures, the membrane is more fluid than in colder temperatures. When the membrane is becoming more fluid and needs to become more stabilized, it will make longer fatty acid chains or saturated fatty acid chains in order to help stabilize the membrane.

Bacteria are also surrounded by a cell wall composed of peptidoglycan (amino acids and sugars). Some eukaryotic cells also have cell walls, but none that are made of peptidoglycan. The outer membrane of gram negative bacteria is rich in lipopolysaccharides, which are combined poly- or oligosaccharide and carbohydrate lipid regions that stimulate the cell's natural immunity.

The outer membrane can bleb out into periplasmic protrusions under stress conditions or upon virulence requirements while encountering a host target cell, and thus such blebs may work as virulence organelles. Bacterial cells provide numerous examples of the diverse ways in which prokaryotic cell membranes are adapted with structures that suit the organism's niche. For example, proteins on the surface of certain bacterial cells aid in their gliding motion. Many gram-negative bacteria have cell membranes which contain ATP-driven protein exporting systems.

2.2.5 Structures

Fluid Mosaic Model

According to the fluid mosaic model of S. J. Singer and G. L. Nicolson (1972), which replaced the earlier model of Davson and Danielli, biological membranes can be considered as a two-dimensional liquid in which lipid and protein molecules diffuse more or less easily. Although the lipid bilayers that form the basis of the membranes do indeed form two-dimensional liquids by themselves, the plasma membrane also contains a large quantity of proteins, which provide more structure. Examples of such structures are protein-protein complexes, pickets and fences formed by the actin-based cytoskeleton, and potentially lipid rafts.

Lipid Bilayer

Lipid bilayers form through the process of self-assembly. The cell membrane consists primarily of a thin layer of amphipathic phospholipids that spontaneously arrange so that the hydrophobic "tail" regions are isolated from the surrounding water while the hydrophilic "head" regions interact with the intracellular (cytosolic) and extracellular faces of the resulting bilayer. This forms a continuous, spherical lipid bilayer. Hydrophobic interactions (also known as the hydrophobic effect) are the major driving forces in the formation of lipid bilayers. An increase in interactions between hydrophobic molecules (causing clustering of hydrophobic regions) allows water molecules to bond more freely with each other, increasing the entropy of the system. This complex interaction can include noncovalent interactions such as van der Waals, electrostatic and hydrogen bonds.

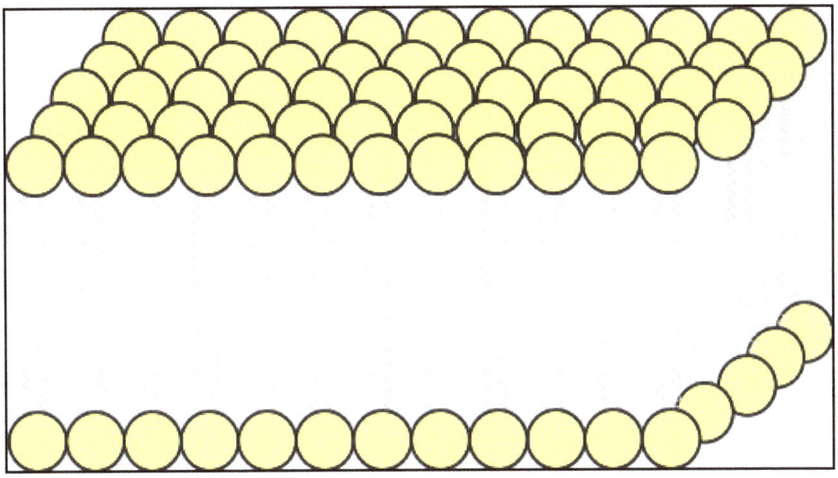

Fig: Diagram of the Arrangement of Amphipathic Lipid Molecules to form a Lipid Bilayer. The Yellow Polar Head Groups Separate the Grey Hydrophobic Tails from the Aqueous Cytosolic and Extracellular Environments

Lipid bilayers are generally impermeable to ions and polar molecules. The arrangement of hydrophilic heads and hydrophobic tails of the lipid bilayer prevent polar solutes (ex. amino acids, nucleic acids, carbohydrates, proteins, and ions) from diffusing across the membrane, but generally allows for the passive diffusion of hydrophobic molecules. This affords the cell the ability to control the movement of these substances via transmembrane protein complexes such as pores, channels and gates. Flippases and scramblases concentrate phosphatidyl serine, which carries a negative charge, on the inner membrane. Along with NANA, this creates an extra barrier to charged moieties moving through the membrane.

Membranes serve diverse functions in eukaryotic and prokaryotic cells. One important role is to regulate the movement of materials into and out of cells. The phospholipid bilayer structure (fluid mosaic model) with specific membrane proteins accounts for the selective permeability of the membrane and passive and active transport mechanisms. In addition, membranes in prokaryotes and in the mitochondria and chloroplasts of eukaryotes facilitate the synthesis of ATP through chemiosmosis.

Membrane Polarity

The apical membrane or luminal membrane of a polarized cell is the surface of the plasma membrane that faces inward to the lumen. This is particularly evident in epithelial and endothelial cells, but also describes other polarized cells, such as neurons. The basolateral membrane or basolateral cell membrane of a polarized cell is the surface of the plasma membrane that forms its basal and lateral surfaces. It faces outwards, towards the interstitium, and away from the lumen. Basolateral membrane is a compound phrase referring to the terms "basal (base) membrane" and "lateral (side) membrane", which, especially in epithelial cells, are identical in composition and activity.

Proteins (such as ion channels and pumps) are free to move from the basal to the lateral surface of the cell or vice versa in accordance with the fluid mosaic model. Tight junctions join epithelial cells near their apical surface to prevent the migration of proteins from the basolateral membrane to the apical membrane. The basal and lateral surfaces thus remain roughly equivalent to one another, yet distinct from the apical surface.

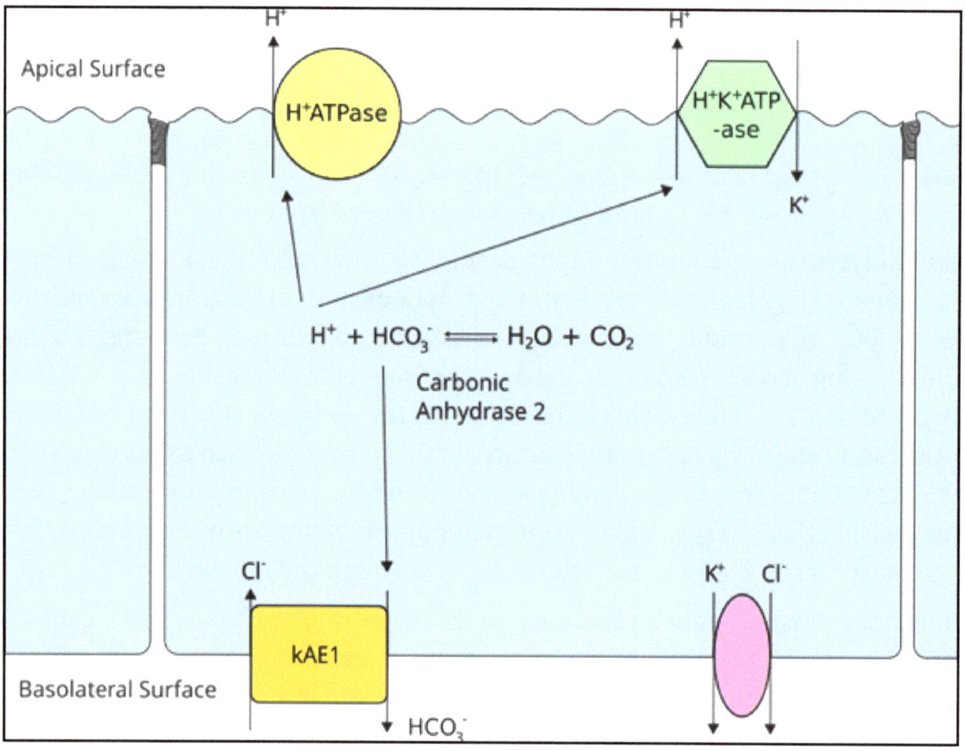

Fig: Alpha Intercalated Cell

Membrane Structures

Cell membrane can form different types of "supramembrane" structures such as caveolae, postsynaptic density, podosomes, invadopodia, focal adhesion, and different types of cell junctions. These structures are usually responsible for cell adhesion, communication, endocytosis and exocytosis. They can be visualized by electron microscopy or fluorescence microscopy. They are composed of specific proteins, such as integrins and cadherins.

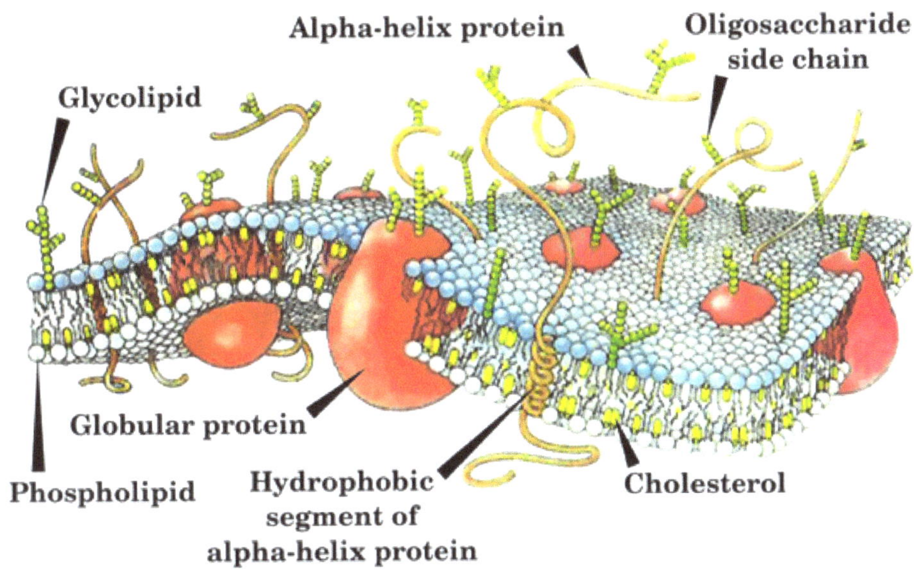

Fig: Diagram of the Cell Membrane's Structures

Cytoskeleton

The cytoskeleton is found underlying the cell membrane in the cytoplasm and provides a scaffolding for membrane proteins to anchor to, as well as forming organelles that extend from the cell. Indeed, cytoskeletal elements interact extensively and intimately with the cell membrane. Anchoring proteins restricts them to a particular cell surface – for example, the apical surface of epithelial cells that line the vertebrate gut – and limits how far they may diffuse within the bilayer. The cytoskeleton is able to form appendage-like organelles, such as cilia, which are microtubule-based extensions covered by the cell membrane, and filopodia, which are actin-based extensions. These extensions are ensheathed in membrane and project from the surface of the cell in order to sense the external environment and/or make contact with the substrate or other cells. The apical surfaces of epithelial cells are dense with actin-based finger-like projections known as microvilli, which increase cell surface area and thereby increase the absorption rate of nutrients. Localized decoupling of the cytoskeleton and cell membrane results in formation of a bleb.

Intracellular Membranes

The content of the cell, inside the cell membrane, is composed of numerous membrane-bound organelles, which contribute to the overall function of the cell. The origin, structure, and function of each organelle leads to a large variation in the cell composition due to the individual uniqueness associated with each organelle.

- Mitochondria and chloroplasts are considered to have evolved from bacteria, known as the endosymbiotic theory. This theory arose from the idea that Paracoccus and Rhodopseudomonas, types of bacteria, share similar functions to mitochondria and blue-green algae (cyanobacteria) share similar functions to chloroplasts. Endosymbiotic theory proposes that through the course of evolution, a eukaryotic cell engulfed these two types of bacteria, leading to the formation of mitochondria and chloroplasts inside eukaryotic cells. This engulfment lead to the double-membranes systems of these organelles in which the outer membrane originated from the host's plasma membrane and the inner membrane was the endosymbiont's plasma membrane. Considering that mitochondria and chloroplasts both contain their own DNA is further support that both of these organelles evolved from engulfed bacteria that thrived inside a eukaryotic cell.
- In eukaryotic cells, the nuclear membrane separates the contents of the nucleus from the cytoplasm of the cell. The nuclear membrane is formed by an inner and outer membrane, providing the strict regulation of materials in to and out of the nucleus. Materials move between the cytosol and the nucleus through nuclear pores in the nuclear membrane. If a cell's nucleus is more active in transcription, its membrane will have more pores. The protein composition of the nucleus can vary greatly from the cytosol as many proteins are unable to cross through pores via diffusion. Within the nuclear membrane, the inner and outer membranes vary in protein composition, and only the outer membrane is continuous with the endoplasmic reticulum (ER) membrane. Like the ER, the outer membrane also possesses ribosomes responsible for producing and transporting proteins into the space between the two membranes. The nuclear membrane disassembles during the early stages of mitosis and reassembles in later stages of mitosis.
- The ER, which is part of the endomembrane system, which makes up a very large portion of the cell's total membrane content. The ER is an enclosed network of tubules and sacs, and its main functions include protein synthesis, and lipid metabolism. There are 2 types of ER, smooth and rough. The rough ER has ribosomes attached to it used for protein synthesis, while the smooth ER is used more for the processing of toxins and calcium regulation in the cell.
- The Golgi apparatus has two interconnected round Golgi cisternae. Compartments of the apparatus forms multiple tubular-reticular

networks responsible for organization, stack connection and cargo transport that display a continuous grape-like stringed vesicles ranging from 50 to 60 nm. The apparatus consists of three main compartments, a flat disc-shaped cisterna with tubular-reticular networks and vesicles.

Variations

The cell membrane has different lipid and protein compositions in distinct types of cells and may have therefore specific names for certain cell types.

- **Sarcolemma in Muscle Cells:** Sarcolemma is the name given to the cell membrane of muscle cells. Although the sarcolemma is similar to other cell membranes, it has other functions that set it apart. For instance, the sarcolemma transmits synaptic signals, helps generate action potentials, and is very involved in muscle contraction. Unlike other cell membranes, the sarcolemma makes up small channels called T-tubules that pass through the entirety of muscle cells. It has also been found that the average sarcolemma is 10 nm thick as opposed to the 4 nm thickness of a general cell membrane.
- **Oolemma is the Cell Membrane in Oocytes:** The oolemma of oocytes, (immature egg cells) are not consistent with a lipid bilayer as they lack a bilayer and do not consist of lipids. Rather, the structure has an inner layer, the fertilization envelope, and the exterior is made up of the vitelline layer, which is made up of glycoproteins; however, channels and proteins are still present for their functions in the membrane.
- **Axolemma:** The specialized plasma membrane on the axons of nerve cells that is responsible for the generation of the action potential. It consists of a granular, densely packed lipid bilayer that works closely with the cytoskeleton components spectrin and actin. These cytoskeleton components are able to bind to and interact with transmembrane proteins in the axolemma.

2.2.6 Permeability

The permeability of a membrane is the rate of passive diffusion of molecules through the membrane. These molecules are known as permeant molecules. Permeability depends mainly on the electric charge and polarity of the molecule and to a lesser extent the molar mass of the molecule.

Due to the cell membrane's hydrophobic nature, small electrically neutral molecules pass through the membrane more easily than charged, large ones. The inability of charged molecules to pass through the cell membrane results in pH partition of substances throughout the fluid compartments of the body.

2.3 LYSIS

Lysis is the breaking down of the membrane of a cell, often by viral, enzymic, or osmotic mechanisms that compromise its integrity. A fluid containing the contents of lysed cells is called a lysate. In molecular biology, biochemistry, and cell biology laboratories, cell cultures may be subjected to lysis in the process of purifying their components, as in protein purification, DNA extraction, RNA extraction, or in purifying organelles.

Many species of bacteria are subject to lysis by the enzyme lysozyme, found in animal saliva, egg white, and other secretions. Phage lytic enzymes (lysins) produced during bacteriophage infection are responsible for the ability of these viruses to lyse bacterial cells. Penicillin and related β-lactam antibiotics cause the death of bacteria through enzyme-mediated lysis that occurs after the drug causes the bacterium to form a defective cell wall. If the cell wall is completely lost and the penicillin was used on gram-positive bacteria, then the bacterium is referred to as a protoplast, but if penicillin was used on gram-negative bacteria, then it is called a spheroplast.

2.3.1 Cytolysis

Cytolysis occurs when a cell bursts due to an osmotic imbalance that has caused excess water to move into the cell. Cytolysis can be prevented by several different mechanisms, including the contractile vacuole that exists in some paramecia, which rapidly pump water out of the cell. Cytolysis does not occur under normal conditions in plant cells because plant cells have a strong cell wall that contains the osmotic pressure, or turgor pressure, that would otherwise cause cytolysis to occur.

- **Oncolysis:** Oncolysis is the destruction of neoplastic cells or of a tumour. The term is also used to refer to the reduction of any swelling.
- **Plasmolysis:** Plasmolysis is the contraction of cells within plants due to the loss of water through osmosis. In a hypertonic environment, the cell membrane peels off the cell wall and the vacuole collapses. These cells will eventually wilt and die unless the flow of water caused by osmosis can stop the contraction of the cell membrane.
- **Immune Response:** Erythrocytes' hemoglobin release free radicals in response to pathogens when lysed by them. This can damage the pathogens.

2.3.2 Applications

Cell lysis is used in laboratories to break open cells and purify or further study their contents. Lysis in the laboratory may be affected by enzymes or detergents or other chaotropic agents. Mechanical disruption of cell membranes, as by repeated freezing and thawing, sonication, pressure, or filtration may also be referred to as lysis. Many laboratory experiments are sensitive to the choice of lysis mechanism; often it is desirable to avoid mechanical shear forces that would denature or degrade sensitive macromolecules, such as proteins and DNA, and different types of detergents can yield different results. The unprocessed solution immediately after lysis but before any further extraction steps is often referred to as a crude lysate.

For example, lysis is used in western and Southern blotting to analyze the composition of specific proteins, lipids, and nucleic acids individually or as complexes. Depending on the detergent used, either all or some membranes are lysed. For example, if only the cell membrane is lysed then gradient centrifugation can be used to collect certain organelles. Lysis is also used for protein purification, DNA extraction, and RNA extraction.

2.3.3 Methods

Several methods for cell lysis exist, sometimes used in combination. Examples include liquid homogenization, freeze thawing, and physical disruption such as sonication, or the use of hypotonic solutions that cause osmotic swelling and eventual bursting of the cell.

- **Chemical Lysis:** This method uses chemical disruption. It is the most popular and simple approach. Chemical lysis chemically deteriorates/ solubilizes the proteins and lipids present within the membrane of targeted cells. Common lysis buffers contain sodium hydroxide (NaOH) and sodium dodecyl sulfate (SDS). Cell lysis is best done at a pH range of 11.5–12.5. Although simple, it is a slow process, taking anywhere from 6 to 12 hours.
- **Acoustic Lysis:** This method uses ultrasonic waves to generate areas of high and low pressure which causes cavitation and in turn, cell lysis. Though this method usually comes out clean, it fails to be cost effective and consistent.
- **Mechanical Lysis:** This method uses physical penetration to pierce or cut a cell membrane.
- **Enzymatic Lysis:** This method uses enzymes such as lysozyme or proteases to disintegrate the cell membrane.

2.4 BACTERIOPHAGE

A bacteriophage also known informally as a phage is a virus that infects and replicates within bacteria and archaea. Their genomes may encode as few as four genes (e.g. MS2) and as many as hundreds of genes. Phages replicate within the bacterium following the injection of their genome into its cytoplasm.

Bacteriophages are among the most common and diverse entities in the biosphere. Bacteriophages are ubiquitous viruses, found wherever bacteria exist. It is estimated there are more than 10^{31} bacteriophages on the planet, more than every other organism on Earth, including bacteria, combined. Viruses are the most abundant biological entity in the water column of the world's oceans, and the second largest component of biomass after prokaryotes, where up to $9x10^8$ virions per millilitre have been found in microbial mats at the surface, and up to 70% of marine bacteria may be infected by bacteriophages.

Bacteriophages were used from the 1920s as an alternative to antibiotics in the former Soviet Union and Central Europe, as well as in France and Brazil. They are seen as a possible therapy against multi-drug-resistant strains of many bacteria.

Bacteriophages are known to interact with the immune system both indirectly via bacterial expression of phage-encoded proteins and directly by influencing innate immunity and bacterial clearance. Phage–host interactions are becoming increasingly important areas of research.

2.4.1 Classification

Bacteriophages occur abundantly in the biosphere, with different genomes and lifestyles. Phages are classified by the International Committee on Taxonomy of Viruses (ICTV) according to morphology and nucleic acid.

The International Committee on Taxonomy of Viruses (ICTV) is the globally recognized authority responsible for the classification and nomenclature of viruses. ICTV develops and maintains a universal taxonomy that organizes viruses based on a combination of genetic, structural, and biological characteristics. Within this framework, prokaryotic viruses—those that infect bacteria and archaea—are systematically categorized to reflect their evolutionary relationships and functional traits.

Prokaryotic viruses, commonly known as bacteriophages (or simply phages when infecting bacteria) and archaeal viruses (when infecting archaea), are highly diverse in structure, genome composition, and replication strategies. These viruses play a crucial role in microbial ecology, gene transfer, and biotechnology. Understanding their classification is essential for studies in virology, microbiology, and biotechnology.

This realm includes double-stranded DNA viruses that typically have icosahedral heads and tails. Most tailed bacteriophages, historically grouped in the order Caudovirales, are now distributed across multiple families and orders in this realm due to advances in genome sequencing and phylogenetic analysis.

Table: ICTV Classification of Prokaryotic (Bacterial and Archaeal) Viruses

Order	Family	Morphology	Nucleic acid	Examples
Belfryvirales	*Turriviridae*	Enveloped, isometric	Linear dsDNA	
Caudovirales	*Ackermannviridae*	Nonenveloped, contractile tail	Linear dsDNA	
	Autographiviridae	Nonenveloped, noncontractile tail (short)	Linear dsDNA	
	Chaseviridae		Linear dsDNA	
	Demerecviridae		Linear dsDNA	
	Drexlerviridae		Linear dsDNA	
	Guenliviridae		Linear dsDNA	
	Herelleviridae	Nonenveloped, contractile tail	Linear dsDNA	
	Myoviridae	Nonenveloped, contractile tail	Linear dsDNA	T4, Mu, P1, P2
	Siphoviridae	Nonenveloped, noncontractile tail (long)	Linear dsDNA	λ, T5, HK97, N15
	Podoviridae	Nonenveloped, noncontractile tail (short)	Linear dsDNA	T7, T3, Φ29, P22
	Rountreeviridae		Linear dsDNA	
	Salasmaviridae		Linear dsDNA	
	Schitoviridae		Linear dsDNA	
	Zobellviridae		Linear dsDNA	
Halopanivirales	*Sphaerolipoviridae*	Enveloped, isometric	Linear dsDNA	
	Simuloviridae	Enveloped, isometric	Linear dsDNA	
	Matshushitaviridae	Enveloped, isometric	Linear dsDNA	
Haloruvirales	*Pleolipoviridae*	Enveloped, pleomorphic	Circular ssDNA, circular dsDNA, or linear dsDNA	
Kalamavirales	*Tectiviridae*	Nonenveloped, isometric	Linear dsDNA	
Lautamovirales	*Finnlakeviridae*	Nonenveloped, isometric	Circular ssDNA	FLiP

Ligamenvirales	*Lipothrixviridae*	Enveloped, rod-shaped	Linear dsDNA	Acidianus filamentous virus 1
	Rudiviridae	Nonenveloped, rod-shaped	Linear dsDNA	Sulfolobus islandicus rod-shaped virus 1
Mindivirales	*Cystoviridae*	Enveloped, spherical	Linear dsRNA	Φ6
Norzivirales	*Atkinsviridae*	Nonenveloped, isometric	Linear ssRNA	
	Duinviridae	Nonenveloped, isometric	Linear ssRNA	
	Fiersviridae	Nonenveloped, isometric	Linear ssRNA	MS2, Qβ
	Solspiviridae	Nonenveloped, isometric	Linear ssRNA	
Petitvirales	*Microviridae*	Nonenveloped, isometric	Circular ssDNA	ΦX174
Primavirales	*Tristromaviridae*	Enveloped, rod-shaped	Linear dsDNA	
Timlovirales	*Blumeviridae*	Nonenveloped, isometric	Linear ssRNA	
	Steitzviridae	Nonenveloped, isometric	Linear ssRNA	
Tubulavirales	*Inoviridae*	Nonenveloped, filamentous	Circular ssDNA	M13
	Paulinoviridae	Nonenveloped, filamentous	Circular ssDNA	
	Plectroviridae	Nonenveloped, filamentous	Circular ssDNA	
Vinavirales	*Corticoviridae*	Nonenveloped, isometric	Circular dsDNA	PM2
Durnavirales	*Picobirnaviridae* (proposal)	Nonenveloped, isometric	Linear dsRNA	

Unassigned	*Ampullaviridae*	Enveloped, bottle-shaped	Linear dsDNA	
	Autolykiviridae	Nonenveloped, isometric	Linear dsDNA	
	Bicaudaviridae	Nonenveloped, lemon-shaped	Circular dsDNA	
	Clavaviridae	Nonenveloped, rod-shaped	Circular dsDNA	
	Fuselloviridae	Nonenveloped, lemon-shaped	Circular dsDNA	Alphafus ellovirus
	Globuloviridae	Enveloped, isometric	Linear dsDNA	
	Guttaviridae	Nonenveloped, ovoid	Circular dsDNA	
	Halspiviridae	Nonenveloped, lemon-shaped	Linear dsDNA	
	Plasmaviridae	Enveloped, pleomorphic	Circular dsDNA	
	Portogloboviridae	Enveloped, isometric	Circular dsDNA	
	Thaspiviridae	Nonenveloped, lemon-shaped	Linear dsDNA	
	Spiraviridae	Nonenveloped, rod-shaped	Circular ssDNA	

It has been suggested that members of Picobirnaviridae infect bacteria, but not mammals.

2.4.2 History

In 1896, Ernest Hanbury Hankin reported that something in the waters of the Ganges and Yamuna rivers in India had a marked antibacterial action against cholera and it could pass through a very fine porcelain filter. In 1915, British bacteriologist Frederick Twort, superintendent of the Brown Institution of London, discovered a small agent that infected and killed bacteria. He believed the agent must be one of the following:

- A stage in the life cycle of the bacteria,
- An enzyme produced by the bacteria themselves, or
- A virus that grew on and destroyed the bacteria.

Twort's research was interrupted by the onset of World War I, as well as a shortage of funding and the discoveries of antibiotics.

Independently, French-Canadian microbiologist Félix d'Hérelle, working at the Pasteur Institute in Paris, announced on 3 September 1917 that he had discovered "an invisible, antagonistic microbe of the dysentery bacillus". For d'Hérelle, there was no question as to the nature of his discovery: "In a flash I had understood: what caused my clear spots was in fact an invisible microbe... a virus parasitic on bacteria." D'Hérelle called the virus a bacteriophage, a bacterium-eater (from the Greek phagein, meaning "to devour"). He also recorded a dramatic account of a man suffering from dysentery who was restored to good health by the bacteriophages. It was d'Hérelle who conducted much research into bacteriophages and introduced the concept of phage therapy. In 1919, in Paris, France, d'Hérelle conducted the first clinical application of a bacteriophage, with the first reported use in the United States being in 1922.

2.4.3 Uses

Phage Therapy

Phages were discovered to be antibacterial agents and were used in the former Soviet Republic of Georgia (pioneered there by Giorgi Eliava with help from the co-discoverer of bacteriophages, Félix d'Hérelle) during the 1920s and 1930s for treating bacterial infections.

D'Herelle "quickly learned that bacteriophages are found wherever bacteria thrive: in sewers, in rivers that catch waste runoff from pipes, and in the stools of convalescent patients."

They had widespread use, including treatment of soldiers in the Red Army. However, they were abandoned for general use in the West for several reasons:

- Antibiotics were discovered and marketed widely. They were easier to make, store, and prescribe.
- Medical trials of phages were carried out, but a basic lack of understanding of phages raised questions about the validity of these trials.
- Publication of research in the Soviet Union was mainly in the Russian or Georgian languages and for many years was not followed internationally.
- The Soviet technology was widely discouraged and in some cases illegal due to the red scare.

The use of phages has continued since the end of the Cold War in Russia, Georgia, and elsewhere in Central and Eastern Europe. The first regulated, randomized, double-blind clinical trial was reported in the Journal of Wound Care in June 2009, which evaluated the safety and efficacy of a bacteriophage cocktail to treat infected venous ulcers of the leg in human patients. The FDA approved the study as a Phase I clinical trial. The study's results demonstrated the safety of therapeutic application of bacteriophages, but did not show efficacy. The authors explained that the use of certain chemicals that are part of standard wound care (e.g. lactoferrin or silver) may have interfered with bacteriophage viability. Shortly after that, another controlled clinical trial in Western Europe (treatment of ear infections caused by Pseudomonas aeruginosa) was reported in the journal Clinical Otolaryngology in August 2009. The study concludes that bacteriophage preparations were safe and effective for treatment of chronic ear infections in humans. Additionally, there have been numerous animal and other experimental clinical trials evaluating the efficacy of bacteriophages for various diseases, such as infected burns and wounds, and cystic fibrosis-associated lung infections, among others. On the other hand, phages of Inoviridae have been shown to complicate biofilms involved in pneumonia and cystic fibrosis and to shelter the bacteria from drugs meant to eradicate disease, thus promoting persistent infection.

Meanwhile, bacteriophage researchers have been developing engineered viruses to overcome antibiotic resistance, and engineering the phage genes responsible for coding enzymes that degrade the biofilm matrix, phage structural proteins, and the enzymes responsible for lysis of the bacterial cell wall. There have been results showing that T4 phages that are small in size and short-tailed can be helpful in detecting E. coli in the human body.

Therapeutic efficacy of a phage cocktail was evaluated in a mouse model with nasal infection of multi-drug-resistant (MDR) A. baumannii. Mice treated with the phage cocktail showed a 2.3-fold higher survival rate compared to those untreated at seven days post-infection.

In 2017, a 68-year-old diabetic patient with necrotizing pancreatitis complicated by a pseudocyst infected with MDR A. baumannii strains was being treated with a cocktail of Azithromycin, Rifampicin, and Colistin for 4 months without results and overall rapidly declining health.

Because discussion had begun of the clinical futility of further treatment, an Emergency Investigational New Drug (eIND) was filed as a last effort to at the very least gain valuable medical data from the situation, and approved, so he was subjected to phage therapy using a percutaneously (PC) injected cocktail containing nine different phages that had been identified as effective against the primary infection strain by rapid isolation and testing techniques (a process which took under a day).

This proved effective for a very brief period, although the patient remained unresponsive and his health continued to worsen; soon isolates of a strain of A. baumannii were being collected from drainage of the cyst that showed resistance to this cocktail, and a second cocktail which was tested to be effective against this new strain was added, this time by intravenous (IV) injection as it had become clear that the infection was more pervasive than originally thought.

Once on the combination of the IV and PC therapy the patient's downward clinical trajectory reversed, and within two days he had awoken from his coma and become responsive. As his immune system began to function he had to be temporarily removed from the cocktail because his fever was spiking to over 104 °F (40 °C), but after two days the phage cocktails were re-introduced at levels he was able to tolerate. The original three-antibiotic cocktail was replaced by minocycline after the bacterial strain was found not to be resistant to this and he rapidly regained full lucidity, although he was not discharged from the hospital until roughly 145 days after phage therapy began. Towards the end of the therapy it was discovered that the bacteria had become resistant to both of the original phage cocktails, but they were continued because they seemed to be preventing minocycline resistance from developing in the bacterial samples collected so were having a useful synergistic effect.

Other

Food Industry

Phages have increasingly been used to safen food products and to forestall spoilage bacteria. Since 2006, the United States Food and Drug Administration (FDA) and United States Department of Agriculture (USDA) have approved several bacteriophage products. LMP-102 (Intralytix) was approved for treating ready-to-eat (RTE) poultry and meat products. In that same year, the FDA approved LISTEX (developed and produced by Micreos) using bacteriophages on cheese to kill Listeria monocytogenes bacteria, in order to give them generally recognized as safe (GRAS) status. In July 2007, the same bacteriophage were approved for use on all food products. In 2011 USDA confirmed that LISTEX is a clean label processing aid and is included in USDA. Research in the field of food safety is continuing to see if lytic phages are a viable option to control other food-borne pathogens in various food products.

Water Indicators

Bacteriophages, including those specific to Escherichia coli, have been employed as indicators of fecal contamination in water sources. Due to their shared structural and biological characteristics, coliphages can serve as proxies for viral fecal contamination and the presence of pathogenic viruses such as rotavirus, norovirus, and HAV. Research conducted on wastewater treatment systems has revealed significant disparities in the behavior of coliphages compared to fecal coliforms, demonstrating a distinct correlation with the recovery of pathogenic viruses at the treatment's conclusion. Establishing a secure discharge threshold, studies have determined that discharges below 3000 PFU/100 mL are considered safe in terms of limiting the release of pathogenic viruses.

Diagnostics

In 2011, the FDA cleared the first bacteriophage-based product for in vitro diagnostic use. The KeyPath MRSA/MSSA Blood Culture Test uses a cocktail of bacteriophage to detect Staphylococcus aureus in positive blood cultures and determine methicillin resistance or susceptibility. The test returns results in about five hours, compared to two to three days for standard microbial identification and susceptibility test methods. It was the first accelerated antibiotic-susceptibility test approved by the FDA.

Counteracting Bioweapons and Toxins

Government agencies in the West have for several years been looking to Georgia and the former Soviet Union for help with exploiting phages for counteracting bioweapons and toxins, such as anthrax and botulism. Developments are continuing among research groups in the U.S. Other uses include spray application in horticulture for protecting plants and vegetable produce from decay and the spread of bacterial disease. Other applications for bacteriophages are as biocides for environmental surfaces, e.g., in hospitals, and as preventative treatments for catheters and medical devices before use in clinical settings. The technology for phages to be applied to dry surfaces, e.g., uniforms, curtains, or even sutures for surgery now exists. Clinical trials reported in Clinical Otolaryngology show success in veterinary treatment of pet dogs with otitis.

Bacterium Sensing and Identification

The sensing of phage-triggered ion cascades (SEPTIC) bacterium sensing and identification method uses the ion emission and its dynamics during phage infection and offers high specificity and speed for detection.

Phage Display

Phage display is a different use of phages involving a library of phages with a variable peptide linked to a surface protein. Each phage genome encodes the variant of the protein displayed on its surface (hence the name), providing a link between the peptide variant and its encoding gene. Variant phages from the library may be selected through their binding affinity to an immobilized molecule (e.g., botulism toxin) to neutralize it. The bound, selected phages can be multiplied by reinfecting a susceptible bacterial strain.

Antimicrobial Drug Discovery

Phage proteins often have antimicrobial activity and may serve as leads for peptidomimetics, i.e. drugs that mimic peptides. Phage-ligand technology makes use of phage proteins for various applications, such as binding of bacteria and bacterial components (e.g. endotoxin) and lysis of bacteria.

2.4.4 Detriments

Dairy Industry

Bacteriophages present in the environment can cause cheese to not ferment. In order to avoid this, mixed-strain starter cultures and culture rotation regimes can be used. Genetic engineering of culture microbes - especially Lactococcus lactis and Streptococcus thermophilus - have been studied for genetic analysis and modification to improve phage resistance. This has especially focused on plasmid and recombinant chromosomal modifications.

Some research has focused on the potential of bacteriophages as antimicrobial against foodborne pathogens and biofilm formation within the dairy industry. As the spread of antibiotic resistance is a main concern within the dairy industry, phages can serve as a promising alternative.

2.4.5 Replication

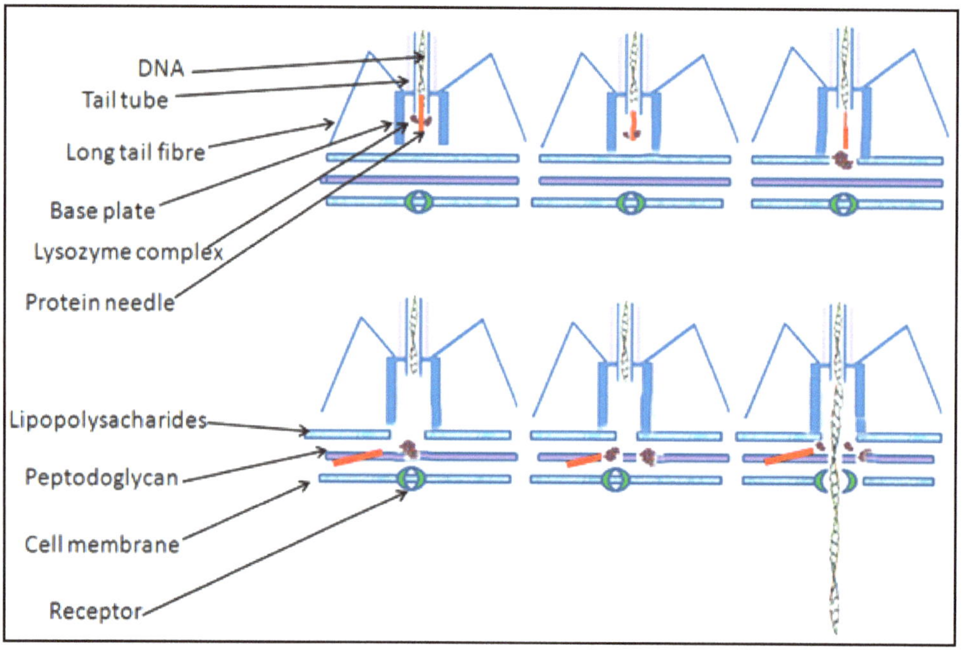

Fig: Diagram of the DNA Injection Process

The life cycle of bacteriophages tends to be either a lytic cycle or a lysogenic cycle. In addition, some phages display pseudolysogenic behaviors.

With lytic phages such as the T4 phage, bacterial cells are broken open (lysed) and destroyed after immediate replication of the virion. As soon as the cell is destroyed, the phage progeny can find new hosts to infect. Lytic phages are more suitable for phage therapy. Some lytic phages undergo a phenomenon known as lysis inhibition, where completed phage progeny will not immediately lyse out of the cell if extracellular phage concentrations are high. This mechanism is not identical to that of the temperate phage going dormant and usually is temporary.

In contrast, the lysogenic cycle does not result in immediate lysing of the host cell. Those phages able to undergo lysogeny are known as temperate phages. Their viral genome will integrate with host DNA and replicate along with it, relatively harmlessly, or may even become established as a plasmid. The virus remains dormant until host conditions deteriorate, perhaps due to depletion of nutrients, then, the endogenous phages (known as prophages) become active. At this point they initiate the reproductive cycle, resulting in lysis of the host cell. As the lysogenic cycle allows the host cell to continue to survive and reproduce, the virus is replicated in all offspring of the cell. An example of a bacteriophage known to follow the lysogenic cycle and the lytic cycle is the phage lambda of E. coli.

Sometimes prophages may provide benefits to the host bacterium while they are dormant by adding new functions to the bacterial genome, in a phenomenon called lysogenic conversion. Examples are the conversion of harmless strains of Corynebacterium diphtheriae or Vibrio cholerae by bacteriophages to highly virulent ones that cause diphtheria or cholera, respectively. Strategies to combat certain bacterial infections by targeting these toxin-encoding prophages have been proposed.

Attachment and Penetration

Bacterial cells are protected by a cell wall of polysaccharides, which are important virulence factors protecting bacterial cells against both immune host defenses and antibiotics. To enter a host cell, bacteriophages bind to specific receptors on the surface of bacteria, including lipopolysaccharides, teichoic acids, proteins, or even flagella. This specificity means a bacteriophage can infect only certain bacteria bearing receptors to which they can bind, which in turn, determines the phage's host range. Polysaccharide-degrading enzymes are virion-associated proteins that enzymatically degrade the capsular outer layer of their hosts at the initial step of a tightly programmed phage infection process. Host growth conditions also influence the ability of the phage to attach and invade them. As phage virions do not move independently, they must rely on random encounters with the correct receptors when in solution, such as blood, lymphatic circulation, irrigation, soil water, etc.

Myovirus bacteriophages use a hypodermic syringe-like motion to inject their genetic material into the cell. After contacting the appropriate receptor, the tail fibers flex to bring the base plate closer to the surface of the cell. This is known as reversible binding. Once attached completely, irreversible binding is initiated and the tail contracts, possibly with the help of ATP present in the tail, injecting genetic material through the bacterial membrane. The injection is accomplished through a sort of bending motion in the shaft by going to the side, contracting closer to the cell and pushing back up. Podoviruses lack an elongated tail sheath like that of a myovirus, so instead, they use their small, tooth-like tail fibers enzymatically to degrade a portion of the cell membrane before inserting their genetic material.

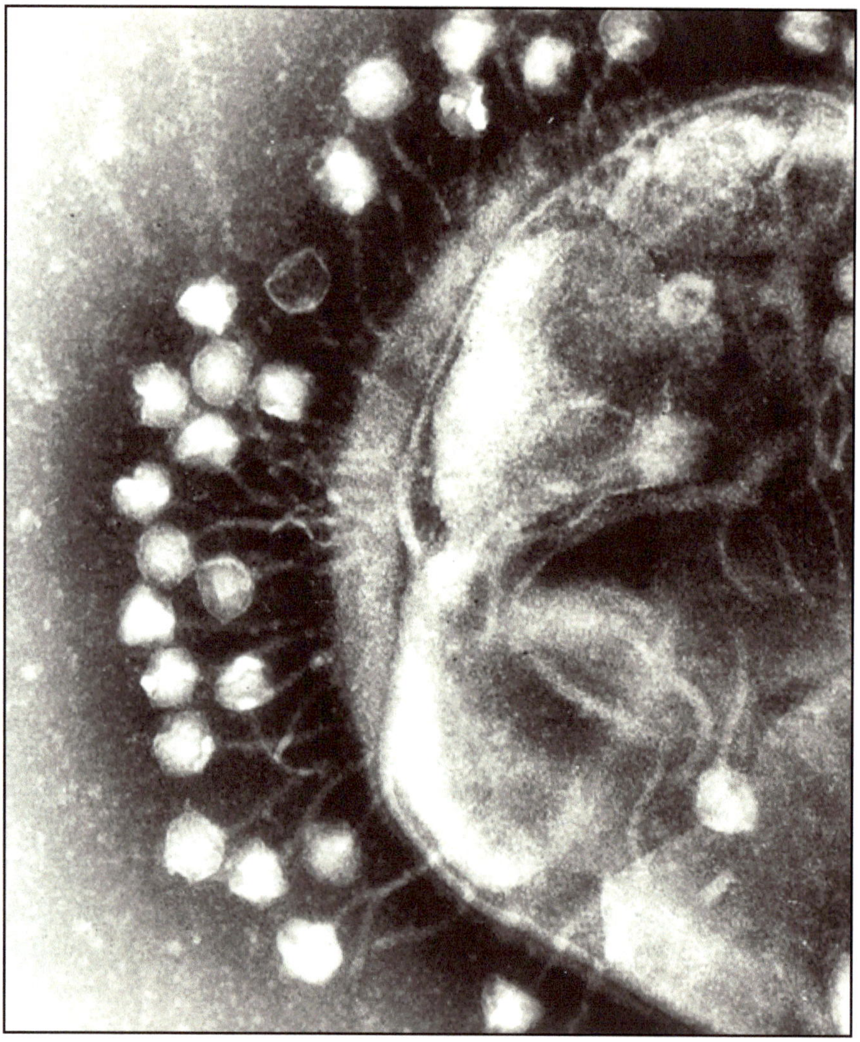

Fig: In this Electron Micrograph of Bacteriophages Attached to a Bacterial Cell, the Viruses are the Size and Shape of Coliphage T1

Synthesis of Proteins and Nucleic Acid

Within minutes, bacterial ribosomes start translating viral mRNA into protein. For RNA-based phages, RNA replicase is synthesized early in the process. Proteins modify the bacterial RNA polymerase so it preferentially transcribes viral mRNA. The host's normal synthesis of proteins and nucleic acids is disrupted, and it is forced to manufacture viral products instead. These products go on to become part of new virions within the cell, helper proteins that contribute to the assemblage of new virions, or proteins involved in cell lysis. In 1972, Walter Fiers (University of Ghent, Belgium) was the first to establish the complete nucleotide sequence of a gene and in 1976, of the viral genome of bacteriophage MS2. Some dsDNA bacteriophages encode ribosomal proteins, which are thought to modulate protein translation during phage infection.

Virion Assembly

In the case of the T4 phage, the construction of new virus particles involves the assistance of helper proteins that act catalytically during phage morphogenesis. The base plates are assembled first, with the tails being built upon them afterward. The head capsids, constructed separately, will spontaneously assemble with the tails.

During assembly of the phage T4 virion, the morphogenetic proteins encoded by the phage genes interact with each other in a characteristic sequence. Maintaining an appropriate balance in the amounts of each of these proteins produced during viral infection appears to be critical for normal phage T4 morphogenesis. The DNA is packed efficiently within the heads. The whole process takes about 15 minutes.

Early studies of bactioriophage T4 (1962-1964) provided an opportunity to gain understanding of virtually all of the genes that are essential for growth of the bacteriophage under laboratory conditions. These studies were made possible by the availability of two classes of conditional lethal mutants. One class of such mutants was referred to as amber mutants. The other class of conditional lethal mutants was referred to as temperature-sensitive mutants Studies of these two classes of mutants led to considerable insight into the functions and interactions of the proteins employed in the machinery of DNA replication, repair and recombination, and on how viruses are assembled from protein and nucleic acid components (molecular morphogenesis).

Release of Virions

Phages may be released via cell lysis, by extrusion, or, in a few cases, by budding. Lysis, by tailed phages, is achieved by an enzyme called endolysin, which attacks and breaks down the cell wall peptidoglycan. An altogether different phage type, the filamentous phage, makes the host cell continually secrete new virus particles. Released virions are described as free, and, unless defective, are capable of infecting a new bacterium. Budding is associated with certain Mycoplasma phages. In contrast to virion release, phages displaying a lysogenic cycle do not kill the host and instead become long-term residents as prophages.

Communication

Research in 2017 revealed that the bacteriophage Φ3T makes a short viral protein that signals other bacteriophages to lie dormant instead of killing the host bacterium. Arbitrium is the name given to this protein by the researchers who discovered it.

2.4.6 Genome Structure

Given the millions of different phages in the environment, phage genomes come in a variety of forms and sizes. RNA phages such as MS2 have the smallest genomes, with only a few kilobases. However, some DNA phages such as T4 may have large genomes with hundreds of genes; the size and shape of the capsid varies along with the size of the genome. The largest bacteriophage genomes reach a size of 735 kb.

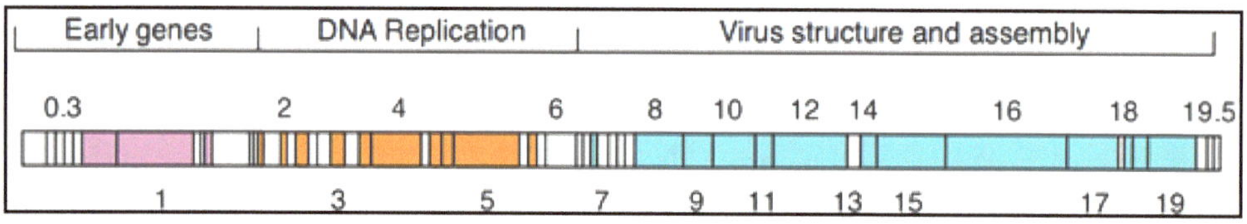

Fig: Schematic View of the 44 kb T7 Phage Genome. Each Box is a Gene. Numbers Indicate Genes (or Rather Open Reading Frames). The "Early", "Middle" (DNA Replication), and "Late" Genes (Virus Structure), Roughly Represent the Time Course of Gene Expression

Bacteriophage genomes can be highly mosaic, i.e. the genome of many phage species appear to be composed of numerous individual modules. These modules may be found in other phage species in different arrangements. Mycobacteriophages, bacteriophages with mycobacterial hosts, have provided excellent examples of this mosaicism. In these mycobacteriophages, genetic assortment may be the result of repeated instances of site-specific recombination and illegitimate recombination (the result of phage genome acquisition of bacterial host genetic sequences). Evolutionary mechanisms shaping the genomes of bacterial viruses vary between different families and depend upon the type of the nucleic acid, characteristics of the virion structure, as well as the mode of the viral life cycle.

Some marine roseobacter phages, also known as roseophages, contain deoxyuridine (dU) instead of deoxythymidine (dT) in their genomic DNA. There is some evidence that this unusual component is a mechanism to evade bacterial defense mechanisms such as restriction endonucleases and CRISPR/Cas systems which evolved to recognize and cleave sequences within invading phages, thereby inactivating them. Other phages have long been known to use unusual nucleotides. In 1963, Takahashi and Marmur identified a Bacillus phage that has dU substituting dT in its genome, and in 1977, Kirnos et al. identified a cyanophage containing 2-aminoadenine (Z) instead of adenine (A).

2.4.7 Systems Biology

The field of systems biology investigates the complex networks of interactions within an organism, usually using computational tools and modeling. For example, a phage genome that enters into a bacterial host cell may express hundreds of phage proteins which will affect the expression of numerous host genes or the host's metabolism. All of these complex interactions can be described and simulated in computer models.

For instance, infection of Pseudomonas aeruginosa by the temperate phage PaP3 changed the expression of 38% (2160/5633) of its host's genes. Many of these effects are probably indirect, hence the challenge becomes to identify the direct interactions among bacteria and phage.

Several attempts have been made to map protein–protein interactions among phage and their host. For instance, bacteriophage lambda was found to interact with its host, E. coli, by dozens of interactions. Again, the significance of many of these interactions remains unclear, but these studies suggest that there most likely are several key interactions and many indirect interactions whose role remains uncharacterized.

2.4.8 Host Resistance

Bacteriophages are a major threat to bacteria and prokaryotes have evolved numerous mechanisms to block infection or to block the replication of bacteriophages within host cells. The CRISPR system is one such mechanism as are retrons and the anti-toxin system encoded by them. The Thoeris defense system is known to deploy a unique strategy for bacterial antiphage resistance via NAD+ degradation.

2.4.9 Bacteriophage–Host Symbiosis

Temperate phages are bacteriophages that integrate their genetic material into the host as extrachromosomal episomes or as a prophage during a lysogenic cycle. Some temperate phages can confer fitness advantages to their host in numerous ways, including giving antibiotic resistance through the transfer or introduction of antibiotic resistance genes (ARGs), protecting hosts from phagocytosis, protecting hosts from secondary infection through superinfection exclusion, enhancing host pathogenicity, or enhancing bacterial metabolism or growth. Bacteriophage–host symbiosis may benefit bacteria by providing selective advantages while passively replicating the phage genome.

2.4.10 In the Environment

Metagenomics has allowed the in-water detection of bacteriophages that was not possible previously. Also, bacteriophages have been used in hydrological tracing and modelling in river systems, especially where surface water and groundwater interactions occur. The use of phages is preferred to the more conventional dye marker because they are significantly less absorbed when passing through ground waters and they are readily detected at very low concentrations. Non-polluted water may contain approximately 2×10^8 bacteriophages per ml.

Bacteriophages are thought to contribute extensively to horizontal gene transfer in natural environments, principally via transduction, but also via transformation. Metagenomics-based studies also have revealed that viromes from a variety of environments harbor antibiotic-resistance genes, including those that could confer multidrug resistance. Recent findings have mapped the complex and intertwined arsenal of anti-phage defense tools in environmental bacteria.

2.4.11 In Humans

Although phages do not infect humans, there are countless phage particles in the human body, given the extensive human microbiome. One's phage population has been called the human phageome, including the "healthy gut phageome" (HGP) and the "diseased human phageome" (DHP). The active phageome of a healthy human (i.e., actively replicating as opposed to nonreplicating, integrated prophage) has been estimated to comprise dozens to thousands of different viruses. There is evidence that bacteriophages and bacteria interact in the human gut microbiome both antagonistically and beneficially.

Preliminary studies have indicated that common bacteriophages are found in 62% of healthy individuals on average, while their prevalence was reduced by 42% and 54% on average in patients with ulcerative colitis (UC) and Crohn's disease (CD). Abundance of phages may also decline in the elderly.

The most common phages in the human intestine, found worldwide, are crAssphages. CrAssphages are transmitted from mother to child soon after birth, and there is some evidence suggesting that they may be transmitted locally. Each person develops their own unique crAssphage clusters. CrAss-like phages also may be present in primates besides humans.

CHAPTER

Filtration

2

Filtration is a physical separation process that separates solid matter and fluid from a mixture using a filter medium that has a complex structure through which only the fluid can pass. Solid particles that cannot pass through the filter medium are described as oversize and the fluid that passes through is called the filtrate. Oversize particles may form a filter cake on top of the filter and may also block the filter lattice, preventing the fluid phase from crossing the filter, known as blinding. The size of the largest particles that can successfully pass through a filter is called the effective pore size of that filter. The separation of solid and fluid is imperfect; solids will be contaminated with some fluid and filtrate will contain fine particles (depending on the pore size, filter thickness and biological activity). Filtration occurs both in nature and in engineered systems; there are biological, geological, and industrial forms. In everyday usage the verb “strain” is more often used; for example, using a colander to drain cooking water from cooked pasta.

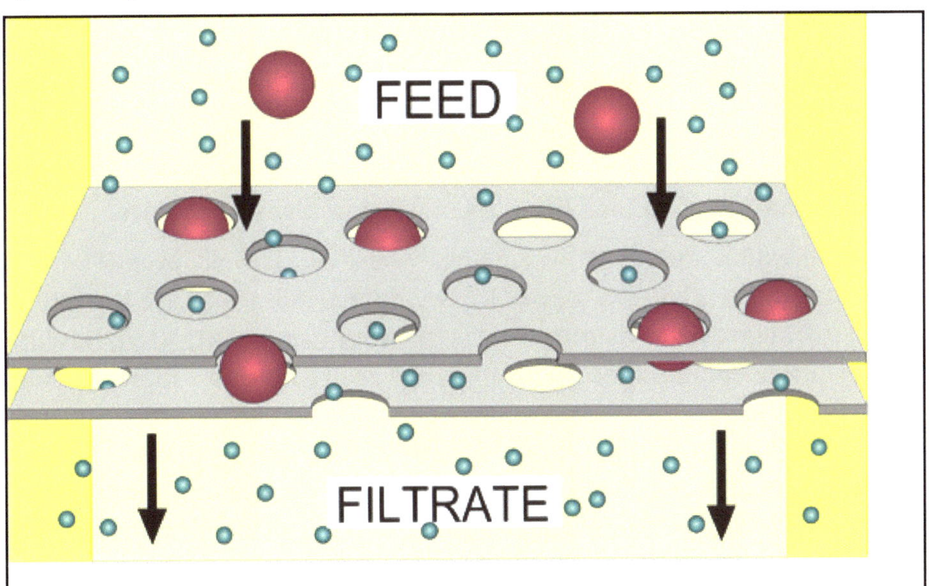

Fig: Diagram of Simple Filtration: Oversize Particles in the Feed cannot Pass through the Lattice Structure of the Filter, while Fluid and Small Particles Pass through, becoming Filtrate

Oil filtration refers to the method of purifying oil by removing impurities that can degrade its quality. Contaminants can enter the oil through various means, including wear and tear of machinery components, environmental factors, and improper handling during oil changes. The primary goal of oil filtration is to enhance the oil's performance, thereby protecting the machinery and extending its service life.

Filtration is also used to describe biological and physical systems that not only separate solids from a fluid stream but also remove chemical species and biological organisms by entrainment, phagocytosis, adsorption and absorption. Examples include slow sand filters and trickling filters. It is also used as a general term for macrophage in which organisms use a variety of means to filter small food particles from their environment. Examples range from the microscopic Vorticella up to the basking shark, one of the largest fishes, and the baleen whales, all of which are described as filter feeders.

3.1 PHYSICAL PROCESSES

- Filtration is used to separate particles and fluid in a suspension, where the fluid can be a liquid, a gas or a supercritical fluid. Depending on the application, either one or both of the components may be isolated.
- Filtration, as a physical operation enables materials of different chemical compositions to be separated. A solvent is chosen which dissolves one component, while not dissolving the other. By dissolving the mixture in the chosen solvent, one component will go into the solution and pass through the filter, while the other will be retained.
- Filtration is widely used in chemical engineering. It may be combined with other unit operations to process the feed stream, as in the biofilter, which is a combined filter and biological digestion device.
- Filtration differs from sieving, where separation occurs at a single perforated layer (a sieve). In sieving, particles that are too big to pass through the holes of the sieve are retained. In filtration, a multilayer lattice retains those particles that are unable to follow the tortuous channels of the filter. Oversize particles may form a cake layer on top of the filter and may also block the filter lattice, preventing the fluid phase from crossing the filter (blinding). Commercially, the term filter is applied to membranes where the separation lattice is so thin that the surface becomes the main zone of particle separation, even though these products might be described as sieves.
- Filtration differs from adsorption, where separation relies on surface charge. Some adsorption devices containing activated charcoal and

ion-exchange resin are commercially called filters, although filtration is not their principal mechanical function.

- Filtration differs from removal of magnetic contaminants from fluids with magnets (typically lubrication oil, coolants and fuel oils) because there is no filter medium. Commercial devices called "magnetic filters" are sold, but the name reflects their use, not their mode of operation.
- In biological filters, oversize particulates are trapped and ingested and the resulting metabolites may be released. For example, in animals (including humans), renal filtration removes waste from the blood, and in water treatment and sewage treatment, undesirable constituents are removed by adsorption into a biological film grown on or in the filter medium, as in slow sand filtration.

3.1.1 Methods

Filters may be used for the purpose of removing unwanted liquid from a solid residue, cleaning unwanted solids from a liquid, or simply to separate the solid from the liquid.

There are many different methods of filtration; all aim to attain the separation of substances. Separation is achieved by some form of interaction between the substance or objects to be removed and the filter. The substance that is to pass through the filter must be a fluid, i.e. a liquid or gas. Methods of filtration vary depending on the location of the targeted material, i.e. whether it is dissolved in the fluid phase or suspended as a solid.

Fig: Hot Filtration, Solution Contained in the Erlenmeyer Flask is Heated on a Hot Plate to Prevent Re-Crystallization of Solids in the Flask Itself

There are several laboratory filtration techniques depending on the desired outcome namely, hot, cold and vacuum filtration. Some of the major purposes of obtaining the desired outcome are, for the removal of impurities from a mixture or, for the isolation of solids from a mixture.

Fig: Hot Filtration for the Separation of Solids from a Hot Solution

Hot filtration method is mainly used to separate solids from a hot solution. This is done to prevent crystal formation in the filter funnel and other apparatus that come in contact with the solution. As a result, the apparatus and the solution used are heated to prevent the rapid decrease in temperature which in turn, would lead to the crystallisation of the solids in the funnel and hinder the filtration process. One of the most important measures to prevent the formation of crystals in the funnel and to undergo effective hot filtration is the use stemless filter funnel. Due to the absence of a stem in the filter funnel, there is a decrease in the surface area of contact between the solution and the stem of the filter funnel, hence preventing re-crystallization of solid in the funnel, and adversely affecting the filtration process.

Fig: Cold Filtration, the Ice Bath is Used to Cool Down the Temperature of the Solution before Undergoing the Filtration Process

Cold filtration method is the use of an ice bath to rapidly cool the solution to be crystallized rather than leaving it to cool slowly in the room atmosphere. This technique results in the formation of very small crystals as opposed to getting large crystals by cooling the solution at room temperature.

Vacuum filtration technique is mostly preferred for small batches of solution to dry small crystals quickly. This method requires a Büchner funnel, filter paper of a smaller diameter than the funnel, Büchner flask, and rubber tubing to connect to a vacuum source.

Centrifugal filtration is carried out by rapidly rotating the substance to be filtered. The more dense material is separated from the less dense matter by the horizontal rotation.

Gravity filtration is the process of pouring the mixture from a higher location to a lower one. It is frequently accomplished via simple filtration, which involves placing filter paper in a glass funnel with the liquid passing through by gravity while the insoluble solid particles are caught by the filter paper. Filter cones, fluted filters, or filtering pipets can all be employed, depending on the amount of the substance at hand. Gravity filtration is in widespread everyday use, for example for straining cooking water from food, or removing contaminants from a liquid.

3.1.2 Filtering Force

Only when a driving force is supplied will the fluid to be filtered be able to flow through the filter media. Gravity, centrifugation, applying pressure to the fluid above the filter, applying a vacuum below the filter, or a combination of these factors may all contribute to this force. In both straightforward laboratory filtrations and massive sand-bed filters, gravitational force alone may be utilized. Centrifuges with a bowl holding a porous filter media can be thought of as filters in which a centrifugal force several times stronger than gravity replaces gravitational force. A partial vacuum is typically provided to the container below the filter media when laboratory filtration is challenging to speed up the filtering process. Depending on the type of filter being used, the majority of industrial filtration operations employ pressure or vacuum to speed up filtering and reduce the amount of equipment needed.

3.1.3 Filter Media

Filter media are the materials used to do the separation of materials. Two main types of filter media are employed in laboratories:

- Surface filters are solid sieves with a mesh to trap solid particles, sometimes with the aid of filter paper (e.g. Büchner funnel, belt filter, rotary vacuum-drum filter, cross-flow filters, screen filter).
- Depth filters, beds of granular material which retain the solid particles as they pass (e.g. sand filter).

Surface filters allow the solid particles, i.e. the residue, to be collected intact; depth filters do not. However, the depth filter is less prone to clogging due to the greater surface area where the particles can be trapped. Also, when the solid particles are very fine, it is often cheaper and easier to discard the contaminated granules than to clean the solid sieve. Filter media can be cleaned by rinsing with solvents or detergents or backwashing. Alternatively, in engineering applications, such as swimming pool water treatment plants, they may be cleaned by backwashing. Self-cleaning screen filters utilize point-of-suction backwashing to clean the screen without interrupting system flow.

Achieving Flow through the Filter

Fluids flow through a filter due to a pressure difference—fluid flows from the high-pressure side to the low-pressure side of the filter. The simplest method to achieve this is by gravity which can be seen in the coffeemaker example. In the laboratory, pressure in the form of compressed air on the feed side (or vacuum on the filtrate side) may be applied to make the filtration process faster, though this may lead to clogging or the passage of fine particles. Alternatively, the liquid may flow through the filter by the force exerted by a pump, a method commonly used in industry when a reduced filtration time is important. In this case, the filter need not be mounted vertically.

3.1.4 Filter Aid

Certain filter aids may be used to aid filtration. These are often incompressible diatomaceous earth, or kieselguhr, which is composed primarily of silica. Also used are wood cellulose and other inert porous solids such as the cheaper and safer perlite. Activated carbon is often used in industrial applications that require changes in the filtrate's properties, such as altering colour or odour.

These filter aids can be used in two different ways. They can be used as a precoat before the slurry is filtered. This will prevent gelatinous-type solids from plugging the filter medium and also give a clearer filtrate. They can also be added to the slurry before filtration. This increases the porosity of the cake and reduces the resistance of the cake during filtration. In a rotary filter, the filter aid may be applied as a precoat; subsequently, thin slices of this layer are sliced off with the cake. The use of filter aids is usually limited to cases where the cake is discarded or where the precipitate can be chemically separated from the filter.

3.1.5 Alternatives

Filtration is a more efficient method for the separation of mixtures than decantation but is much more time-consuming. If very small amounts of solution are involved, most of the solution may be soaked up by the filter medium.

An alternative to filtration is centrifugation. Instead of filtering the mixture of solid and liquid particles, the mixture is centrifuged to force the (usually) denser solid to the bottom, where it often forms a firm cake. The liquid above can then be decanted. This method is especially useful for separating solids that do not filter well, such as gelatinous or fine particles. These solids can clog or pass through the filter, respectively.

3.2 BIOLOGICAL FILTRATION

Biological filtration may take place inside an organism, or the biological component may be grown on a medium in the material being filtered. Removal of solids, emulsified components, organic chemicals and ions may be achieved by ingestion and digestion, adsorption or absorption. Because of the complexity of biological interactions, especially in multi-organism communities, it is often not possible to determine which processes are achieving the filtration result. At the molecular level, it may often be by individual catalytic enzyme actions within an individual organism. The waste products of some organisms may subsequently broken down by other organisms to extract as much energy as possible and in so doing reduce complex organic molecules to very simple inorganic species such as water, carbon dioxide and nitrogen.

3.2.1 Excretion

In mammals, reptiles, and birds, the kidneys function by renal filtration whereby the glomerulus selectively removes undesirable constituents such as urea, followed by selective reabsorption of many substances essential for the body to maintain homeostasis. The complete process is termed excretion by urination. Similar but often less complex solutions are deployed in all animals, even the protozoa, where the contractile vacuole provides a similar function.

3.2.2 Biofilms

A biofilm is a syntrophic community of microorganisms in which cells stick to each other and often also to a surface. These adherent cells become embedded within a slimy extracellular matrix that is composed of extracellular polymeric substances (EPSs). The cells within the biofilm produce the EPS components, which are typically a polymeric combination of extracellular polysaccharides, proteins, lipids and DNA. Because they have a three-dimensional structure and represent a community lifestyle for microorganisms, they have been metaphorically described as "cities for microbes".

Biofilms may form on living (biotic) or non-living (abiotic) surfaces and can be common in natural, industrial, and hospital settings. They may constitute a microbiome or be a portion of it. The microbial cells growing in a biofilm are physiologically distinct from planktonic cells of the same organism, which, by contrast, are single cells that may float or swim in a liquid medium. Biofilms can form on the teeth of most animals as dental plaque, where they may cause tooth decay and gum disease.

Microbes form a biofilm in response to a number of different factors, which may include cellular recognition of specific or non-specific attachment sites on a surface, nutritional cues, or in some cases, by exposure of planktonic cells to sub-inhibitory concentrations of antibiotics. A cell that switches to the biofilm mode of growth undergoes a phenotypic shift in behavior in which large suites of genes are differentially regulated.

A biofilm may also be considered a hydrogel, which is a complex polymer that contains many times its dry weight in water. Biofilms are not just bacterial slime layers but biological systems; the bacteria organize themselves into a coordinated functional community. Biofilms can attach to a surface such as a tooth or rock, and may include a single species or a diverse group of microorganisms. Subpopulations of cells within the biofilm differentiate to perform various activities for motility, matrix production, and sporulation, supporting the overall success of the biofilm. The biofilm bacteria can share nutrients and are sheltered from harmful factors in the environment, such as desiccation, antibiotics, and a host body's immune system. A biofilm usually begins to form when a free-swimming, planktonic bacterium attaches to a surface.

Origin and Formation

Origin of Biofilms

Biofilms are thought to have arisen during primitive Earth as a defense mechanism for prokaryotes, as the conditions at that time were too harsh for their survival. They can be found very early in Earth's fossil records (about 3.25 billion years ago) as both Archaea and Bacteria, and commonly protect prokaryotic cells by providing them with homeostasis, encouraging the development of complex interactions between the cells in the biofilm.

Formation of Biofilms

The formation of a biofilm begins with the attachment of free-floating microorganisms to a surface. The first colonist bacteria of a biofilm may adhere to the surface initially by the weak van der Waals forces and hydrophobic effects. If the colonists are not immediately separated from the surface, they can anchor themselves more permanently using cell adhesion structures such as pili. A unique group of Archaea that inhabit anoxic groundwater have similar structures called hami. Each hamus is a long tube with three hook attachments that are used to attach to each other or to a surface, enabling a community to develop. Hyperthermophilic archaeon Pyrobaculum calidifontis produce bundling pili which are homologous to the bacterial TasA filaments, a major component of the extracellular matrix in bacterial biofilms, which contribute to biofilm stability. TasA homologs are encoded by many other archaea, suggesting mechanistic similarities and evolutionary connection between bacterial and archaeal biofilms.

Biofilm of golden hydrophobic bacteria; ceiling of Golden Dome Cave, a lava tube in Lava Beds National Monument Hydrophobicity can also affect the ability of bacteria to form biofilms. Bacteria with increased hydrophobicity have reduced repulsion between the substratum and the bacterium. Some bacteria species are not able to attach to a surface on their own successfully due to their limited motility but are instead able to anchor themselves to the matrix or directly to other, earlier bacteria colonists. Non-motile bacteria cannot recognize surfaces or aggregate together as easily as motile bacteria.

During surface colonization bacteria cells are able to communicate using quorum sensing (QS) products such as N-acyl homoserine lactone (AHL). Once colonization has begun, the biofilm grows by a combination of cell division and recruitment. Polysaccharide matrices typically enclose bacterial biofilms. The matrix exopolysaccharides can trap QS autoinducers within the biofilm to prevent predator detection and ensure bacterial survival. In addition to the polysaccharides, these matrices may also contain material from the surrounding environment, including but not limited to minerals, soil particles, and blood components, such as erythrocytes and fibrin. The final stage of biofilm formation is known as development, and is the stage in which the biofilm is established and may only change in shape and size.

The development of a biofilm may allow for an aggregate cell colony to be increasingly tolerant or resistant to antibiotics. Cell-cell communication or quorum sensing has been shown to be involved in the formation of biofilm in several bacterial species.

Development

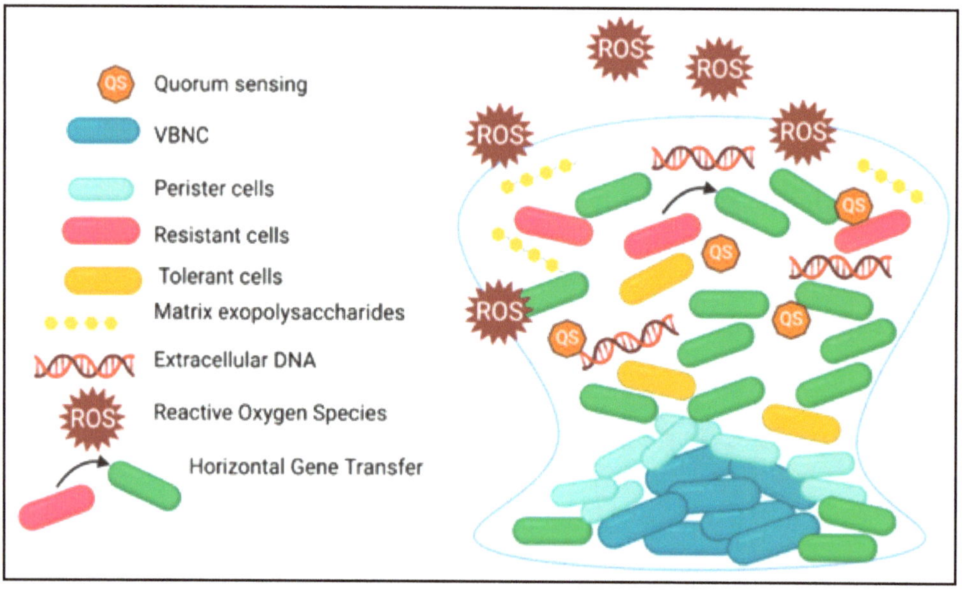

Fig: Mature Biofilm Structure

Biofilm is characterised by heterogenous environment and the presence of a variety of subpopulations. A biofilm structure is composed of metabolically active (both resistant and tolerant) and non-active cells (viable but not culturable cells and persisters) as well as polymer matrix consisting of polysaccharide, extracellular DNA and proteins.

Biofilm growth is associated with an escalated level of mutations and horizontal gene transfer which is promoted in due to the packed and dense structure. Bacteria in biofilms communicate by quorum sensing, which activates genes participating in virulence factors production. Biofilms are the product of a microbial developmental process. The process is summarized by five major stages of biofilm development

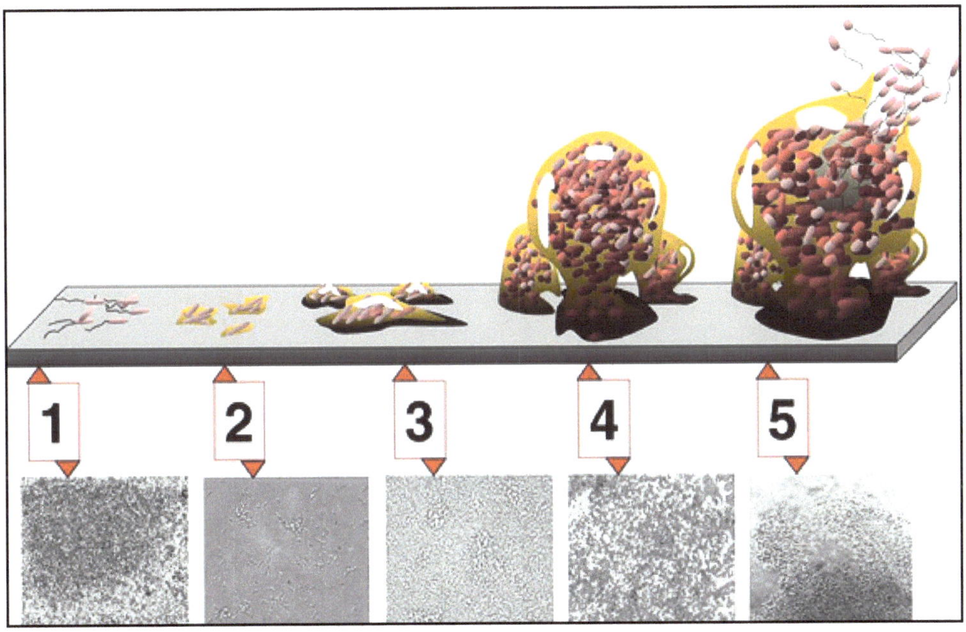

Fig: Five Stages of Biofilm Development: (1) Initial Attachment, (2) Irreversible Attachment, (3) Maturation I, (4) Maturation II, (5) Dispersion

Dispersal

Dispersal of cells from the biofilm colony is an essential stage of the biofilm life cycle. Dispersal enables biofilms to spread and colonize new surfaces. Enzymes that degrade the biofilm extracellular matrix, such as dispersin B and deoxyribonuclease, may contribute to biofilm dispersal. Enzymes that degrade the biofilm matrix may be useful as anti-biofilm agents. Evidence has shown that a fatty acid messenger, cis-2-decenoic acid, is capable of inducing dispersion and inhibiting growth of biofilm colonies. Secreted by Pseudomonas aeruginosa, this compound induces cyclo heteromorphic cells in several species of bacteria and the yeast Candida albicans. Nitric oxide has also been shown to trigger the dispersal of biofilms of several bacteria species at sub-toxic concentrations. Nitric oxide has potential as a treatment for patients that have chronic infections caused by biofilms.

It was generally assumed that cells dispersed from biofilms immediately go into the planktonic growth phase. However, studies have shown that the physiology of dispersed cells from Pseudomonas aeruginosa biofilms is highly different from that of planktonic and biofilm cells. Hence, the dispersal process is a unique stage during the transition from biofilm to planktonic lifestyle in bacteria. Dispersed cells are found to be highly virulent against macrophages and Caenorhabditis elegans, but highly sensitive towards iron stress, as compared with planktonic cells.

Furthermore, Pseudomonas aeruginosa biofilms undergo distinct spatiotemporal dynamics during biofilm dispersal or disassembly, with contrasting consequences in recolonization and disease dissemination. Biofilm dispersal induced bacteria to activate dispersal genes to actively depart from biofilms as single cells at consistent velocities but could not recolonize fresh surfaces.

In contrast, biofilm disassembly by degradation of a biofilm exopolysaccharide released immotile aggregates at high initial velocities, enabling the bacteria to recolonize fresh surfaces and cause infections in the hosts efficiently. Hence, biofilm dispersal is more complex than previously thought, where bacterial populations adopting distinct behavior after biofilm departure may be the key to survival of bacterial species and dissemination of diseases.

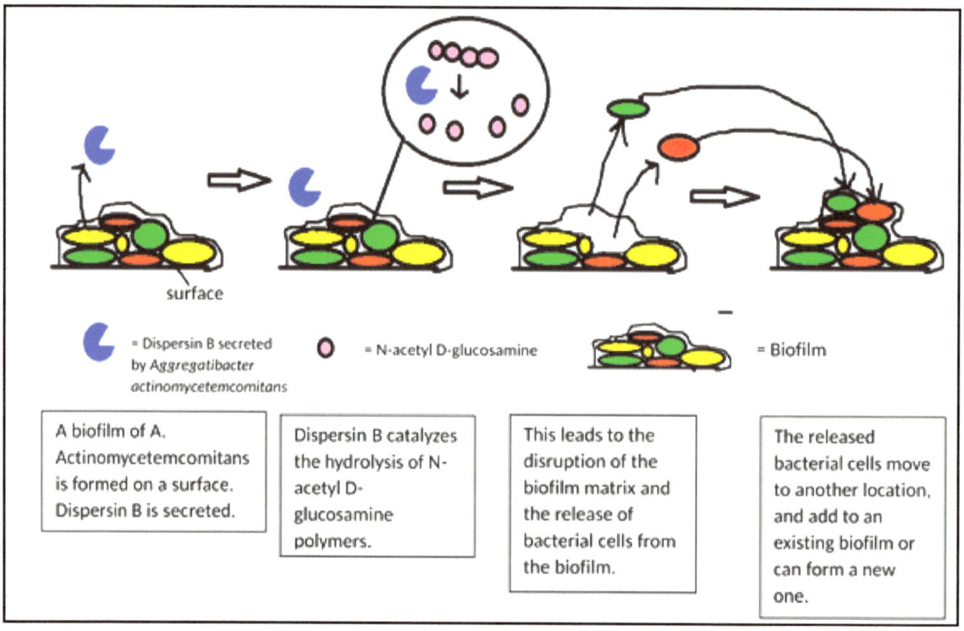

Fig: Biofilm Dispersal

Properties

Biofilms are usually found on solid substrates submerged in or exposed to an aqueous solution, although they can form as floating mats on liquid surfaces and also on the surface of leaves, particularly in high humidity climates. Given sufficient resources for growth, a biofilm will quickly grow to be macroscopic (visible to the naked eye). Biofilms can contain many different types of microorganism, e.g. bacteria, archaea, protozoa, fungi and algae; each group performs specialized metabolic functions. However, some organisms will form single-species films under certain conditions. The social structure (cooperation/competition) within a biofilm depends highly on the different species present.

Extracellular Matrix

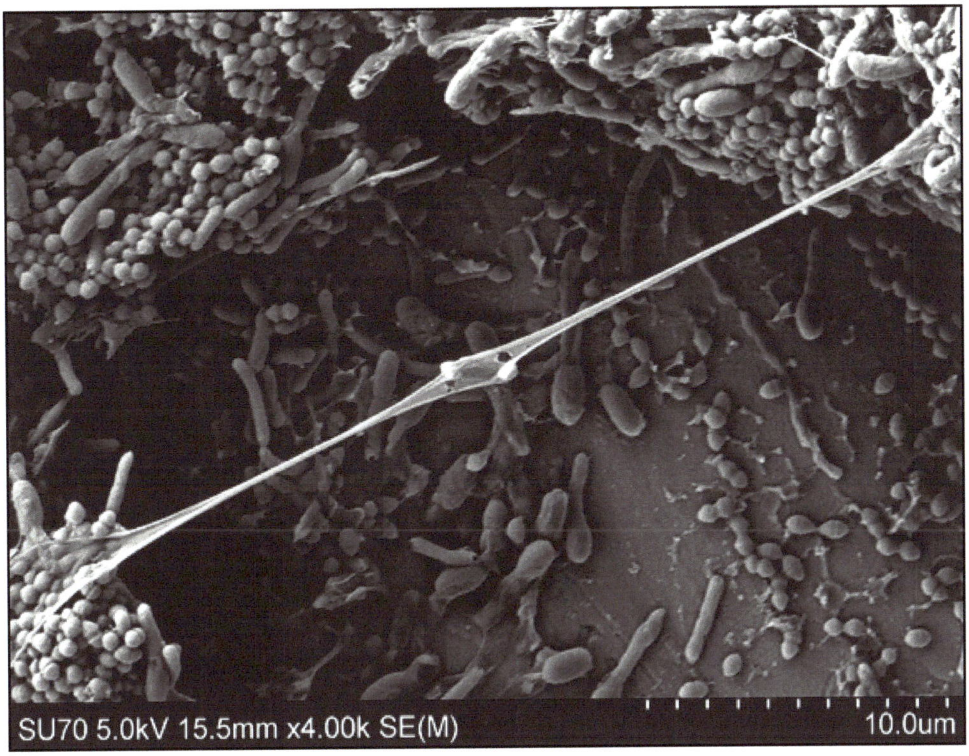

Fig: Scanning Electron Micrograph of Mixed-Culture Biofilm, Demonstrating in Detail a Spatially Heterogeneous Arrangement of Bacterial Cells and Extracellular Polymeric Substances

The EPS matrix consists of exopolysaccharides, proteins and nucleic acids. A large proportion of the EPS is more or less strongly hydrated, however, hydrophobic EPS also occur; one example is cellulose which is produced by a range of microorganisms. This matrix encases the cells within it and facilitates communication among them through biochemical signals as well as gene exchange. The EPS matrix also traps extracellular enzymes and keeps them in close proximity to the cells. Thus, the matrix represents an external digestion system and allows for stable synergistic microconsortia of different species. Some biofilms have been found to contain water channels that help distribute nutrients and signalling molecules. This matrix is strong enough that under certain conditions, biofilms can become fossilized (stromatolites).

Bacteria living in a biofilm usually have significantly different properties from free-floating bacteria of the same species, as the dense and protected environment of the film allows them to cooperate and interact in various ways. One benefit of this environment is increased resistance to detergents and antibiotics, as the dense extracellular matrix and the outer layer of cells protect the interior of the community. In some cases antibiotic resistance can be increased up to 5,000 times. Lateral gene transfer is often facilitated within bacterial and archaeal biofilms and can leads to a more stable biofilm structure. Extracellular DNA is a major structural component of many different microbial biofilms. Enzymatic degradation of extracellular DNA can weaken the biofilm structure and release microbial cells from the surface.

However, biofilms are not always less susceptible to antibiotics. For instance, the biofilm form of Pseudomonas aeruginosa has no greater resistance to antimicrobials than do stationary-phase planktonic cells, although when the biofilm is compared to logarithmic-phase planktonic cells, the biofilm does have greater resistance to antimicrobials. This resistance to antibiotics in both stationary-phase cells and biofilms may be due to the presence of persister cells.

Habitats

Fig: Mats of Bacterial Biofilm Color the Hot Springs in Yellowstone National Park. The Longest Raised Mat Area is about Half a Meter Long

Biofilms are ubiquitous in organic life. Nearly every species of microorganism have mechanisms by which they can adhere to surfaces and to each other. Biofilms will form on virtually every non-shedding surface in non-sterile aqueous or humid environments. Biofilms can grow in the most extreme environments: from, for example, the extremely hot, briny waters of hot springs ranging from very acidic to very alkaline, to frozen glaciers.

Biofilms can be found on rocks and pebbles at the bottoms of most streams or rivers and often form on the surfaces of stagnant pools of water. Biofilms are important components of food chains in rivers and streams and are grazed by the aquatic invertebrates upon which many fish feed. Biofilms are found on the surface of and inside plants. They can either contribute to crop disease or, as in the case of nitrogen-fixing rhizobia on root nodules, exist symbiotically with the plant. Examples of crop diseases related to biofilms include citrus canker, Pierce's disease of grapes, and bacterial spot of plants such as peppers and tomatoes.

Fig: Thermophilic Bacteria in the Outflow of Mickey Hot Springs, Oregon, Approximately 20 mm Thick

Percolating Filters

Percolating filters in sewage treatment works are highly effective removers of pollutants from settled sewage liquor. They work by trickling the liquid over a bed of hard material which is designed to have a very large surface area. A complex biofilm develops on the surface of the medium which absorbs, adsorbs and metabolises the pollutants. The biofilm grows rapidly and when it becomes too thick to retain its grip on the media it washes off and is replaced by newly grown film. The washed off ("sloughed" off) film is settled out of the liquid stream to leave a highly purified effluent.

Slow Sand Filter

Slow sand filters are used in water purification for treating raw water to produce a potable product. They work through the formation of a biofilm called the hypogeal layer or Schmutzdecke in the top few millimetres of the fine sand layer. The

Schmutzdecke is formed in the first 10–20 days of operation and consists of bacteria, fungi, protozoa, rotifera and a range of aquatic insect larvae. As an epigeal biofilm ages, more algae tend to develop and larger aquatic organisms may be present including some bryozoa, snails and annelid worms. The surface biofilm is the layer that provides the effective purification in potable water treatment, the underlying sand providing the support medium for this biological treatment layer. As water passes through the hypogeal layer, particles of foreign matter are trapped in the mucilaginous matrix and soluble organic material is adsorbed. The contaminants are metabolised by the bacteria, fungi and protozoa. The water produced from an exemplary slow sand filter is of excellent quality with 90–99% bacterial cell count reduction.

Rhizosphere

Plant-beneficial microbes can be categorized as plant growth-promoting rhizobacteria. These plant growth-promoters colonize the roots of plants, and provide a wide range of beneficial functions for their host including nitrogen fixation, pathogen suppression, anti-fungal properties, and the breakdown of organic materials. One of these functions is the defense against pathogenic, soil-borne bacteria and fungi by way of induced systemic resistance (ISR) or induced systemic responses triggered by pathogenic microbes (pathogen-induced systemic acquired resistance). Plant exudates act as chemical signals for host specific bacteria to colonize. Rhizobacteria colonization steps include attractions, recognition, adherence, colonization, and growth. Bacteria that have been shown to be beneficial and form biofilms include Bacillus, Pseudomonas, and Azospirillum. Biofilms in the rhizosphere often result in pathogen or plant induced systemic resistances. Molecular properties on the surface of the bacterium cause an immune response in the plant host. These microbe associated molecules interact with receptors on the surface of plant cells, and activate a biochemical response that is thought to include several different genes at a number of loci. Several other signaling molecules have been linked to both induced systemic responses and pathogen-induced systemic responses, such as jasmonic acid and ethylene. Cell envelope components such as bacterial flagella and lipopolysaccharides, which are recognized by plant cells as components of pathogens. Certain iron metabolites produced by Pseudomonas have also been shown to create an induced systemic response. This function of the biofilm helps plants build stronger resistance to pathogens.

Plants that have been colonized by PGPR forming a biofilm have gained systemic resistances and are primed for defense against pathogens. This means that the genes necessary for the production of proteins that work towards defending the plant against pathogens have been expressed, and the plant has a “stockpile” of compounds to release to fight off pathogens. A primed defense system is much faster in responding to pathogen induced infection, and may be able to deflect pathogens before they are able to establish themselves. Plants increase the production of lignin, reinforcing cell walls and making it difficult for pathogens to penetrate into the cell, while also cutting off nutrients to already infected cells, effectively halting the invasion. They produce antimicrobial compounds such as phytoalexins, chitinases, and proteinase inhibitors, which prevent the growth of pathogens. These functions of disease suppression and pathogen resistance ultimately lead to an increase in agricultural production and a decrease in the use of chemical pesticides, herbicides, and fungicides because there is a reduced amount of crop loss due to disease. Induced systemic resistance and pathogen-induced systemic acquired resistance are both potential functions of biofilms in the rhizosphere, and should be taken into consideration when applied to new age agricultural practices because of their effect on disease suppression without the use of dangerous chemicals.

Mammalian Gut

Studies in 2003 discovered that the immune system supports biofilm development in the large intestine. This was supported mainly with the fact that the two most abundantly produced molecules by the immune system also support biofilm production and are associated with the biofilms developed in the gut. This is especially important because the appendix holds a mass amount of these bacterial biofilms. This discovery helps to distinguish the possible function of the appendix and the idea that the appendix can help reinoculate the gut with good gut flora. However, modified or disrupted states of biofilms in the gut have been connected to diseases such as inflammatory bowel disease and colorectal cancer.

Human Environment

In the human environment, biofilms can grow in showers very easily since they provide a moist and warm environment for them to thrive. Mold biofilms on ceilings may form due to roof leaks. They can form inside water and sewage pipes and cause clogging and corrosion. On floors and counters, they can make sanitation difficult in food preparation areas. In soil, they can cause bioclogging. In cooling- or heating-water systems, they are known to reduce heat transfer. Biofilms in marine engineering systems, such as pipelines of the offshore oil and gas industry, can lead to substantial corrosion problems. Corrosion is mainly due to abiotic factors; however, at least 20% of corrosion is caused by microorganisms that are attached to the metal subsurface (i.e., microbially influenced corrosion).

Ship Fouling

Bacterial adhesion to boat hulls serves as the foundation for biofouling of seagoing vessels. Once a film of bacteria forms, it is easier for other marine organisms such as barnacles to attach. Such fouling can reduce maximum vessel speed by up to 20%, prolonging voyages and consuming fuel. Time in dry dock for refitting and repainting reduces the productivity of shipping assets, and the useful life of ships is also reduced due to corrosion and mechanical removal (scraping) of marine organisms from ships' hulls.

Stromatolites

Stromatolites are layered accretionary structures formed in shallow water by the trapping, binding and cementation of sedimentary grains by microbial biofilms, especially of cyanobacteria. Stromatolites include some of the most ancient records of life on Earth, and are still forming today.

Dental Plaque

Within the human body, biofilms are present on the teeth as dental plaque, where they may cause tooth decay and gum disease. These biofilms can either be in an uncalcified state that can be removed by dental instruments, or a calcified state which is more difficult to remove. Removal techniques can also include antimicrobials.

Dental plaque is an oral biofilm that adheres to the teeth and consists of many species of both bacteria and fungi (such as Streptococcus mutans and Candida albicans), embedded in salivary polymers and microbial extracellular products. The accumulation of microorganisms subjects the teeth and gingival tissues to high concentrations of bacterial metabolites which results in dental disease. Biofilm on the surface of teeth is frequently subject to oxidative stress and acid stress. Dietary carbohydrates can cause a dramatic decrease in pH in oral biofilms to values of 4 and below (acid stress). A pH of 4 at body temperature of 37 °C causes depurination of DNA, leaving apurinic (AP) sites in DNA, especially loss of guanine.

Dental plaque biofilm can result in dental caries if it is allowed to develop over time. An ecologic shift away from balanced populations within the dental biofilm is driven by certain (cariogenic) microbiological populations beginning to dominate when the environment favors them. The shift to an acidogenic, aciduric, and cariogenic microbiological population develops and is maintained by frequent consumption of fermentable dietary carbohydrate.

The resulting activity shift in the biofilm (and resulting acid production within the biofilm, at the tooth surface) is associated with an imbalance of demineralization over remineralization, leading to net mineral loss within dental hard tissues (enamel and then dentin), the symptom being a carious lesion, or cavity.

By preventing the dental plaque biofilm from maturing or by returning it back to a non-cariogenic state, dental caries can be prevented and arrested. This can be achieved through the behavioral step of reducing the supply of fermentable carbohydrates (i.e. sugar intake) and frequent removal of the biofilm (i.e., toothbrushing).

Intercellular Communication

A peptide pheromone quorum sensing signaling system in S. mutans includes the competence stimulating peptide (CSP) that controls genetic competence. Genetic competence is the ability of a cell to take up DNA released by another cell. Competence can lead to genetic transformation, a form of sexual interaction, favored under conditions of high cell density and/or stress where there is maximal opportunity for interaction between the competent cell and the DNA released from nearby donor cells. This system is optimally expressed when S. mutans cells reside in an actively growing biofilm. Biofilm grown S. mutans cells are genetically transformed at a rate 10- to 600-fold higher than S. mutans growing as free-floating planktonic cells suspended in liquid.

When the biofilm, containing S. mutans and related oral streptococci, is subjected to acid stress, the competence regulon is induced, leading to resistance to being killed by acid. As pointed out by Michod et al., transformation in bacterial pathogens likely provides for effective and efficient recombinational repair of DNA damages. It appears that S. mutans can survive the frequent acid stress in oral biofilms, in part, through the recombinational repair provided by competence and transformation.

Predator-Prey Interactions

Predator-prey interactions between biofilms and bacterivores, such as the soil-dwelling nematode Caenorhabditis elegans, had been extensively studied. Via the production of sticky matrix and formation of aggregates, Yersinia pestis biofilms can prevent feeding by obstructing the mouth of C. elegans. Moreover, Pseudomonas aeruginosa biofilms can impede the slithering motility of C. elegans, termed as 'quagmire phenotype', resulting in trapping of C. elegans within the biofilms and preventing the exploration of nematodes to feed on susceptible biofilms. This significantly reduced the ability of predator to feed and reproduce, thereby promoting the survival of biofilms. Pseudomonas aeruginosa biofilms can also mask their chemical signatures, where they reduced the diffusion of quorum sensing molecules into the environment and prevented the detection of C. elegans.

Taxonomic Diversity

Many different bacteria form biofilms, including gram-positive (e.g. Bacillus spp, Listeria monocytogenes, Staphylococcus spp, and lactic acid bacteria, including Lactobacillus plantarum and Lactococcus lactis) and gram-negative species (e.g. Escherichia coli, or Pseudomonas aeruginosa). Cyanobacteria also form biofilms in aquatic environments.

Biofilms are formed by bacteria that colonize plants, e.g. Pseudomonas putida, Pseudomonas fluorescens, and related pseudomonads which are common plant-associated bacteria found on leaves, roots, and in the soil, and the majority of their natural isolates form biofilms. Several nitrogen-fixing symbionts of legumes such as Rhizobium leguminosarum and Sinorhizobium meliloti form biofilms on legume roots and other inert surfaces.

Along with bacteria, biofilms are also generated by archaea and by a range of eukaryotic organisms, including fungi e.g. Cryptococcus laurentii and microalgae. Among microalgae, one of the main progenitors of biofilms are diatoms, which colonise both fresh and marine environments worldwide.

Infectious Diseases

Biofilms have been found to be involved in a wide variety of microbial infections in the body, by one estimate 80% of all infections. Infectious processes in which biofilms have been implicated include common problems such as bacterial vaginosis, urinary tract infections, catheter infections, middle-ear infections, formation of dental plaque, gingivitis, coating contact lenses, and less common but more lethal processes such as endocarditis, infections in cystic fibrosis, and infections of permanent indwelling devices such as joint prostheses, heart valves, and intervertebral disc. The first visual evidence of a biofilm was recorded after spine surgery. It was found that in the absence of clinical presentation of infection, impregnated bacteria could form a biofilm around an implant, and this biofilm can remain undetected via contemporary diagnostic methods, including swabbing. Implant biofilm is frequently present in "aseptic" pseudarthrosis cases. Furthermore, it has been noted that bacterial biofilms may impair cutaneous wound healing and reduce topical antibacterial efficiency in healing or treating infected skin wounds. The diversity of P. aeruginosa cells within a biofilm is thought to make it harder to treat the infected lungs of people with cystic fibrosis. Early detection of biofilms in wounds is crucial to successful chronic wound management. Although many techniques have developed to identify planktonic bacteria in viable wounds, few have been able to quickly and accurately identify bacterial biofilms. Future studies are needed to find means of identifying and monitoring biofilm colonization at the bedside to permit timely initiation of treatment.

It has been shown that biofilms are present on the removed tissue of 80% of patients undergoing surgery for chronic sinusitis. The patients with biofilms were shown to have been denuded of cilia and goblet cells, unlike the controls without biofilms who had normal cilia and goblet cell morphology. Biofilms were also found on samples from two of 10 healthy controls mentioned. The species of bacteria from intraoperative cultures did not correspond to the bacteria species in the biofilm on the respective patient's tissue. In other words, the cultures were negative though the bacteria were present. New staining techniques are being developed to differentiate bacterial cells growing in living animals, e.g. from tissues with allergy-inflammations.

Research has shown that sub-therapeutic levels of β-lactam antibiotics induce biofilm formation in Staphylococcus aureus. This sub-therapeutic level of antibiotic may result from the use of antibiotics as growth promoters in agriculture, or during the normal course of antibiotic therapy. The biofilm formation induced by low-level methicillin was inhibited by DNase, suggesting that the sub-therapeutic levels of antibiotic also induce extracellular DNA release. Moreover, from an evolutionary point of view, the creation of the tragedy of the commons in pathogenic microbes may provide advanced therapeutic ways for chronic infections caused by biofilms via genetically engineered invasive cheaters who can invade wild-types 'cooperators' of pathogenic bacteria until cooperator populations go to extinction or overall population 'cooperators and cheaters ' go to extinction.

Pseudomonas Aeruginosa

P. aeruginosa represents a commonly used biofilm model organism since it is involved in different types of biofilm-associated chronic infections. Examples of such infections include chronic wounds, chronic otitis media, chronic prostatitis and chronic lung infections in cystic fibrosis (CF) patients. About 80% of CF patients have chronic lung infection, caused mainly by P. aeruginosa growing in a non-surface attached biofilms surround by PMN. The infection remains present despite aggressive antibiotic therapy and is a common cause of death in CF patients due to constant inflammatory damage to the lungs. In patients with CF, one therapy for treating early biofilm development is to employ DNase to structurally weaken the biofilm.

Biofilm formation of P. aeruginosa, along with other bacteria, is found in 90% of chronic wound infections, which leads to poor healing and high cost of treatment estimated at more than US$25 billion every year in the United States. In order to minimize the P. aeruginosa infection, host epithelial cells secrete antimicrobial peptides, such as lactoferrin, to prevent the formation of the biofilms.

Streptococcus Pneumoniae

Streptococcus pneumoniae is the main cause of community-acquired pneumonia and meningitis in children and the elderly, and of sepsis in HIV-infected persons. When S. pneumoniae grows in biofilms, genes are specifically expressed that respond to oxidative stress and induce competence. Formation of a biofilm depends on competence stimulating peptide (CSP). CSP also functions as a quorum-sensing peptide. It not only induces biofilm formation, but also increases virulence in pneumonia and meningitis.

It has been proposed that competence development and biofilm formation is an adaptation of S. pneumoniae to survive the defenses of the host. In particular, the host's polymorphonuclear leukocytes produce an oxidative burst to defend against the invading bacteria, and this response can kill bacteria by damaging their DNA. Competent S. pneumoniae in a biofilm have the survival advantage that they can more easily take up transforming DNA from nearby cells in the biofilm to use for recombinational repair of oxidative damages in their DNA. Competent S. pneumoniae can also secrete an enzyme (murein hydrolase) that destroys non-competent cells (fratricide) causing DNA to be released into the surrounding medium for potential use by the competent cells.

The insect antimicrobial peptide cecropin A can destroy planktonic and sessile biofilm-forming uropathogenic E. coli cells, either alone or when combined with the antibiotic nalidixic acid, synergistically clearing infection in vivo (in the insect host Galleria mellonella) without off-target cytotoxicity. The multi-target mechanism of action involves outer membrane permeabilization followed by biofilm disruption triggered by the inhibition of efflux pump activity and interactions with extracellular and intracellular nucleic acids.

Escherichia Coli

Escherichia coli biofilms are responsible for many intestinal infectious diseases. The Extraintestinal group of E. coli (ExPEC) is the dominant bacterial group that attacks the urinary system, which leads to urinary tract infections. The biofilm formation of these pathogenic E. coli is hard to eradicate due to the complexity of its aggregation structure, and it has a significant contribution to developing aggressive medical complications, increase in hospitalization rate, and cost of treatment.

The development of E. coli biofilm is a common leading cause of urinary tract infections (UTI) in hospitals through its contribution to developing medical device-associated infections. Catheter-associated urinary tract infections (CAUTI) represent the most common hospital-acquired infection due to the formation of the pathogenic E. coli biofilm inside the catheters.

Staphylococcus Aureus

Staphylococcus aureus pathogen can attack skin and lungs, leading to skin infection and pneumonia. Moreover, the biofilm infections network of S. aureus plays a critical role in preventing immune cells, such as macrophages from eliminating and destroying bacterial cells. Furthermore, biofilm formation by bacteria, such as S. aureus, not only develops resistance against antibiotic medication but also develop internal resistance toward antimicrobial peptides (AMPs), leading to preventing the inhibition of the pathogen and maintaining its survival.

Serratia Marcescens

Serratia marcescens is a fairly common opportunistic pathogen that can form biofilms on various surfaces, including medical devices such as catheters and implants, as well as natural environments like soil and water. The formation of biofilms by S. marcescens is a serious concern because of its ability to adhere to and colonize surfaces, protecting itself from host immune responses and antimicrobial agents. This strength makes infections caused by S. marcescens challenging to treat, specifically in hospitals where the bacterium can cause severe, and specific, infections.

Research suggests that biofilm formation by S. marcescens is a process controlled by both nutrient cues and the quorum-sensing system. Quorum sensing influences the bacterium's ability to adhere to surfaces and establish mature biofilms, whereas the availability of specific nutrients can enhance or inhibit biofilm development.

S. marcescens creates biofilms that ultimately develop into a highly porous, thread-like structure composed of chains of cells, filaments, and cell clusters. Research has shown that S. marcescens biofilms exhibit complex structural organization, including the formation of microcolonies and channels that facilitate nutrient and waste exchange. The production of extracellular polymeric substances (EPS) is a key factor in biofilm development, contributing to the bacterium's adhesion and resistance to antimicrobial agents. In addition to its role in healthcare-associated infections, S. marcescens biofilms have been implicated in the deterioration of industrial equipment and processes. For example, biofilm growth in cooling towers can lead to biofouling and reduced efficiency.

Efforts to control and prevent biofilm formation by S. marcescens involve the use of antimicrobial coatings on medical devices, the development of targeted biofilm disruptors, and improved sterilization protocols. Further research into the molecular mechanisms governing S. marcescens biofilm formation and persistence is crucial for developing effective strategies to combat its associated risks. The use of indole compounds has been studied to be used as protection against biofilm formation.

Uses and Impact

In Medicine

It is suggested that around two-thirds of bacterial infections in humans involve biofilms. Infections associated with the biofilm growth usually are challenging to eradicate. This is mostly due to the fact that mature biofilms display antimicrobial tolerance, and immune response evasions. Biofilms often form on the inert surfaces of implanted devices such as catheters, prosthetic cardiac valves and intrauterine devices. Some of the most difficult infections to treat are those associated with the use of medical devices.

The rapidly expanding worldwide industry for biomedical devices and tissue engineering related products is already at $180 billion per year, yet this industry continues to suffer from microbial colonization. No matter the sophistication, microbial infections can develop on all medical devices and tissue engineering constructs. 60-70% of hospital-acquired infections are associated with the implantation of a biomedical device. This leads to 2 million cases annually in the U.S., costing the healthcare system over $5 billion in additional healthcare expenses.

The level of antibiotic resistance in a biofilm is much greater than that of non-biofilm bacteria, and can be as much as 5,000 times greater. The extracellular matrix of biofilm is considered one of the leading factors that can reduce the penetration of antibiotics into a biofilm structure and contributes to antibiotic resistance. Further, it has been demonstrated that the evolution of resistance to antibiotics may be affected by the biofilm lifestyle. Bacteriophage therapy can disperse the biofilm generated by antibiotic-resistant bacteria.

It has been shown that the introduction of a small current of electricity to the liquid surrounding a biofilm, together with small amounts of antibiotic can reduce the level of antibiotic resistance to levels of non-biofilm bacteria. This is termed the bioelectric effect. The application of a small DC current on its own can cause a biofilm to detach from its surface. A study showed that the type of current used made no difference to the bioelectric effect.

In Industry

Biofilms can also be harnessed for constructive purposes. For example, many sewage treatment plants include a secondary treatment stage in which waste water passes over biofilms grown on filters, which extract and digest organic compounds. In such biofilms, bacteria are mainly responsible for removal of organic matter (BOD), while protozoa and rotifers are mainly responsible for removal of suspended solids (SS), including pathogens and other microorganisms. Slow sand filters rely on biofilm development in the same way to filter surface water from lake, spring or river sources for drinking purposes. What is regarded as clean water is effectively a waste material to these microcellular organisms. Biofilms can help eliminate petroleum oil from contaminated oceans or marine systems. The oil is eliminated by the hydrocarbon-degrading activities of communities of hydrocarbonoclastic bacteria (HCB). Biofilms are used in microbial fuel cells (MFCs) to generate electricity from a variety of starting materials, including complex organic waste and renewable biomass. Biofilms are also relevant for the improvement of metal dissolution in bioleaching industry, and aggregation of microplastics pollutants for convenient removal from the environment.

Food Industry

Biofilms have become problematic in several food industries due to the ability to form on plants and during industrial processes. Bacteria can survive long periods of time in water, animal manure, and soil, causing biofilm formation on plants or in the processing equipment. The buildup of biofilms can affect the heat flow across a surface and increase surface corrosion and frictional resistance of fluids. These can lead to a loss of energy in a system and overall loss of products. Along with economic problems, biofilm formation on food poses a health risk to consumers due to the ability to make the food more resistant to disinfectants As a result, from 1996 to 2010 the Centers for Disease Control and Prevention estimated 48 million foodborne illnesses per year. Biofilms have been connected to about 80% of bacterial infections in the United States.

In produce, microorganisms attach to the surfaces and biofilms develop internally. During the washing process, biofilms resist sanitization and allow bacteria to spread across the produce, especially via kitchen utensils. This problem is also found in ready-to-eat foods, because the foods go through limited cleaning procedures before consumption Due to the perishability of dairy products and limitations in cleaning procedures, resulting in the buildup of bacteria, dairy is susceptible to biofilm formation and contamination. The bacteria can spoil the products more readily and contaminated products pose a health risk to consumers. One species of bacteria that can be found in various industries and is a major cause of foodborne disease is Salmonella. Large amounts of Salmonella contamination can be found in the poultry processing industry as about 50% of Salmonella strains can produce biofilms on poultry farms. Salmonella increases the risk of foodborne illnesses when the poultry products are not cleaned and cooked correctly. Salmonella is also found in the seafood industry where biofilms form from seafood borne pathogens on the seafood itself as well as in water. Shrimp products are commonly affected by Salmonella because of unhygienic processing and handling techniques The preparation practices of shrimp and other seafood products can allow for bacteria buildup on the products.

New forms of cleaning procedures are being tested to reduce biofilm formation in these processes which will lead to safer and more productive food processing industries. These new forms of cleaning procedures also have a profound effect on the environment, often releasing toxic gases into the groundwater reservoirs. As a response to the aggressive methods employed in controlling biofilm formation, there are a number of novel technologies and chemicals under investigation that can prevent either the proliferation or adhesion of biofilm-secreting microbes.

Latest proposed biomolecules presenting marked anti-biofilm activity include a range of metabolites such as bacterial rhamnolipids and even plant- and animal-derived alkaloids.

In Aquaculture

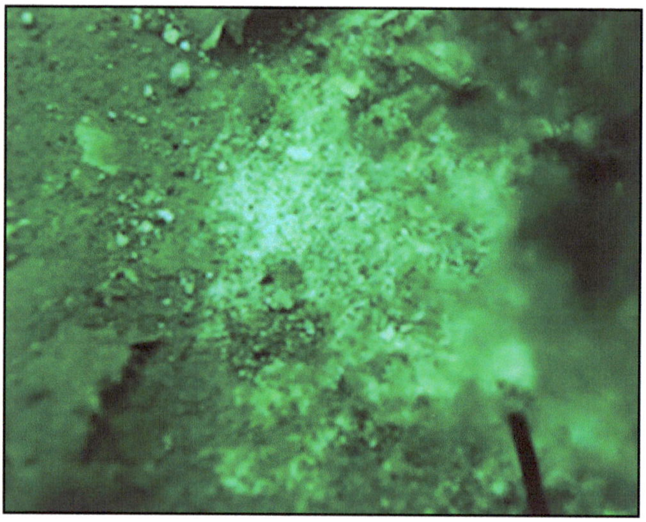

Fig: A Biofilm from the Dead Sea

In shellfish and algal aquaculture, biofouling microbial species tend to block nets and cages and ultimately outcompete the farmed species for space and food. Bacterial biofilms start the colonization process by creating microenvironments that are more favorable for biofouling species. In the marine environment, biofilms could reduce the hydrodynamic efficiency of ships and propellers, lead to pipeline blockage and sensor malfunction, and increase the weight of appliances deployed in seawater. Numerous studies have shown that biofilm can be a reservoir for potentially pathogenic bacteria in freshwater aquaculture. Moreover, biofilms are important in establishing infections on the fish.

As mentioned previously, biofilms can be difficult to eliminate even when antibiotics or chemicals are used in high doses. The role that biofilm plays as reservoirs of bacterial fish pathogens has not been explored in detail but it certainly deserves to be studied.

Eukaryotic

Along with bacteria, biofilms are often initiated and produced by eukaryotic microbes. The biofilms produced by eukaryotes is usually occupied by bacteria and other eukaryotes alike, however the surface is cultivated and EPS is secreted initially by the eukaryote. Both fungi and microalgae are known to form biofilms in such a way. Biofilms of fungal origin are important aspects of human infection and fungal pathogenicity, as the fungal infection is more resistant to antifungals.

In the environment, fungal biofilms are an area of ongoing research. One key area of research is fungal biofilms on plants. For example, in the soil, plant associated fungi including mycorrhiza have been shown to decompose organic matter and protect plants from bacterial pathogens.

Biofilms in aquatic environments are often founded by diatoms. The exact purpose of these biofilms is unknown, however there is evidence that the EPS produced by diatoms facilitates both cold and salinity stress. These eukaryotes interact with a diverse range of other organisms within a region known as the phycosphere, but importantly are the bacteria associated with diatoms, as it has been shown that although diatoms excrete EPS, they only do so when interacting with certain bacteria species.

Horizontal Gene Transfer

Horizontal gene transfer is the lateral transfer of genetic material between cellular organisms. It happens frequently in prokaryotes, and less frequently in eukaryotes. In bacteria, horizontal gene transfer can occur through transformation (uptake of free floating DNA in the environment), transduction (virus mediated DNA uptake), or conjugation (transfer of DNA between pili structures of two adjacent bacteria). Recent studies have also uncovered other mechanisms, such as membrane vesicle transmission or gene transfer agents. Biofilms promote horizontal gene transfer in a variety of ways.

Bacterial conjugation has been shown to accelerate biofilm formation in difficult environment due to the robust connections established by the conjugative pili. These connections can often foster cross-species transfer events due to the diverse heterogeneity of many biofilms. Additionally, biofilms are structurally confined by a polysaccharide matrix, providing the close spatial requirements for conjugation. Transformation is also frequently observed in biofilms.

Bacterial autolysis is a key mechanism in biofilm structural regulation, providing an abundant source of competent DNA primed for transformative uptake. In some instances, inter-biofilm quorum sensing can enhance the competence of free floating eDNA, further promoting transformation. Stx gene transfer through bacteriophage carriers has been witnessed within biofilms, which suggests that biofilms are also a suitable environment for transduction.

Membrane vesicles HGT occurs when released membrane vesicles (containing genetic information) fuse with a recipient bacteria, and release genetic material into the bacteria's cytoplasm. Recent research has revealed that membrane vesicle HGT can promote single-strain biofilm formation, yet the role membrane vesicle HGT plays in the formation of multistrain biofilms is still unknown. GTAs, or gene transfer agents, are phage-like particles produced by the host bacteria and contain random DNA fragments from the host bacteria genome. HGT within biofilms can confer antibiotic resistance or increased pathogenicity across the biofilms' population, promoting biofilm homeostasis.

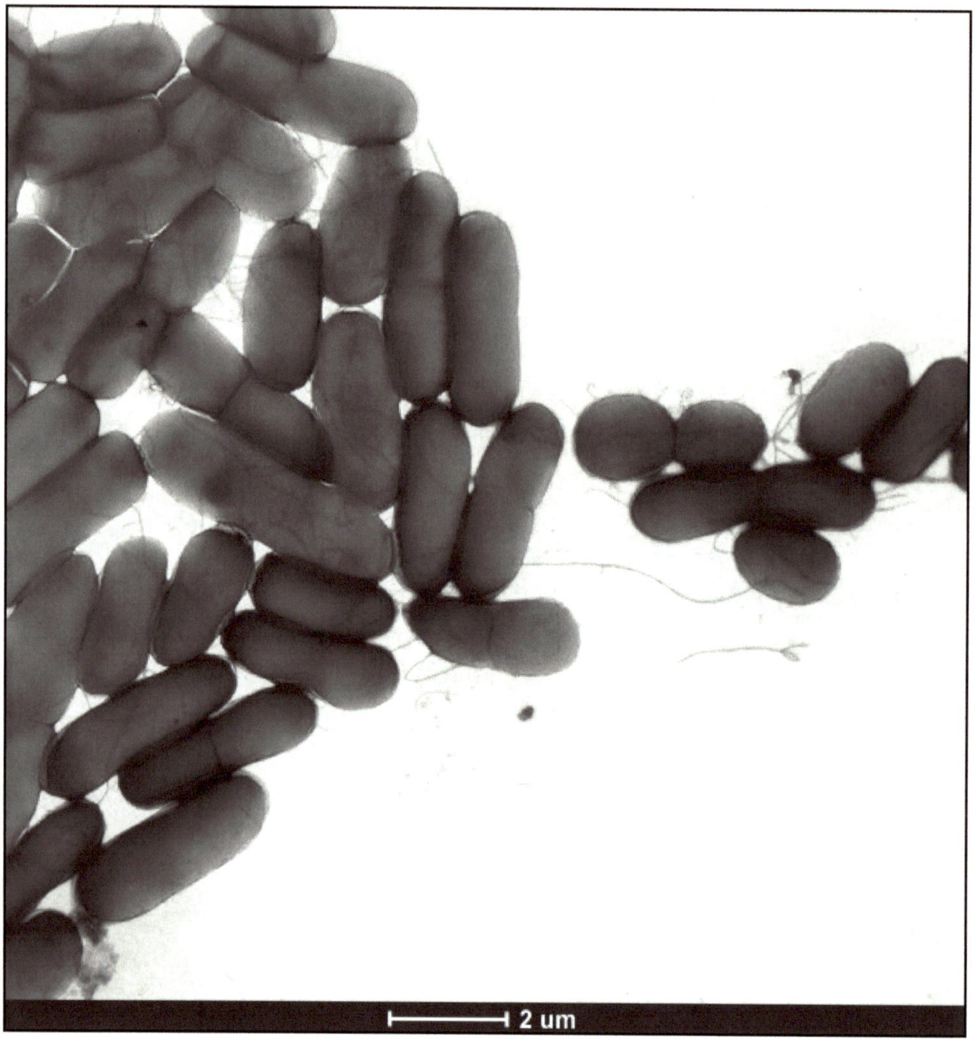

Fig: Transmission Electron Micrograph Showing Bacteria Escherichia Coli that form Extensive Biofilms using a Network of Conjugative F-Pili

Examples

Conjugative plasmids may encode biofilm-associated proteins, such as PtgA, PrgB, or PrgC which promote cell adhesion (required for early biofilm formation). Genes encoding type III fimbriae are found in pOLA52 (Klebsiella pneumoniae plasmid) which promote conjugative-pilus-dependent biofilm formation.

Transformation commonly occurs within biofilms. A phenomenon called fratricide can be seen among streptococcal species in which cell-wall degrading enzymes are released, lysing neighboring bacteria and releasing their DNA. This DNA can then be taken up by the surviving bacteria (transformation). Competence stimulating peptides may play an important role in biofilm formation among S. pneumoniae and S. mutans as well. Among V. cholerae, the competence pilus itself promotes cell aggregation through pilus-pilus interactions at the beginning of biofilm formation.

Phage invasion may play a role in biofilm life cycles, lysing bacteria and releasing their eDNA, which strengthens biofilm structures and can be taken up by neighboring bacteria in transformation. Biofilm destruction caused by the E. coli phage Rac and the P. aeruginosa prophage Pf4 causes detachment of cells from the biofilm. Detachment is a biofilm phenomenon which requires more study, but is hypothesized to proliferate the bacterial species that comprise the biofilm.

Membrane vesicle HGT has been witnessed occurring in marine environments, among Neisseria gonorrhoeae, Pseudomonas aeruginosa, Helicobacter pylori, and among many other bacterial species. Even though membrane vesicle HGT has been shown as a contributing factor in biofilm formation, research is still required to prove that membrane vesicle mediated HGT occurs within biofilms. Membrane vesicle HGT has also been shown to modulate phage-bacteria interactions in Bacillus subtilis SPP1 phage-resistant cells (lacking the SPP1 receptor protein). Upon exposure to vesicles containing receptors, transduction of pBT163 (a cat-encoding plasmid) occurs, resulting in the expression of the SPP1 receptor protein, opening the receptive bacteria to future phage infection.

Recent research has shown that the archaeal species H. volcanii has some biofilm phenotypes similar to bacterial biofilms such as differentiation and HGT, which required cell-cell contact and involved formation of cytosolic bridges and cellular fusion events.

Cultivation Devices

There is a wide variety of biofilm cultivation devices to mimic natural or industrial environments. Although it is important to consider that the particular experimental platform for biofilm research determines what kind of biofilm is cultivated and the data that can be extracted. These devices can be grouped into the following:

- Microtiter plate (MTP) systems and MBEC Assay® [formerly the Calgary Biofilm Device (CBD)].
- BioFilm Ring Test Error in Webarchive template: Empty url. (BRT) or clinical Biofilm Ring Test (cBRT).
- Robbins Device or modified Robbins Device (such as the MPMR-10PMMA or the Bio-inLine Biofilm Reactor).
- Drip Flow Biofilm Reactor®.

- Rotary devices (such as the CDC Biofilm Reactor®, the Rotating Disk Reactor, the Biofilm Annular Reactor, the Industrial Surfaces Biofilm Reactor, or the Constant Depth Film Fermenter).
- Flow chambers or flow cells (such as the Coupon Evaluation Flow Cell, Transmission Flow Cell, and Capillary Flow Cell from BioSurface Technologies).
- Microfluidic approaches, such as 3D-bacterial "biofilm-dispersal-then-recolonization" (BDR) microfluidic model.

3.2.3 Filter Feeders

Filter feeders are organisms that obtain their food by filtering their, generally aquatic, environment. Many of the protozoa are filter feeders using a range of adaptations including rigid spikes of protoplasm held in the water flow as in the suctoria to various arrangements of beating cillia to direct particles to the mouth including organisms such as Vorticella which have a complex ring of cilia which create a vortex in the flow drafting particles into the oral cavity. Similar feeding techniques are used by the Rotifera and the Ectoprocta. Many aquatic arthropods are filter feeders. Some use rhythmical beating of abdominal limbs to create a water current to the mouth whilst the hairs on the legs trap any particle. Others such as some caddis flies spin fine webs in the water flow to trap particles.

3.3 EXAMPLES

Many filtration processes include more than one filtration mechanism, and particulates are often removed from the fluid first to prevent clogging of downstream elements.

Particulate filtration includes:

- The coffee filter to separate the coffee infusion from the grounds.
- HEPA filters in air conditioning to remove particles from air.
- Belt filters to extract precious metals in mining.
- Vertical plate filter such as those used in Merrill–Crowe process.
- The Nutsche filter is typically used in pharmaceutical applications or batch processes that need to capture solids.
- Furnaces use filtration to prevent the furnace elements from fouling with particulates.
- A pneumatic conveying system such as an industrial exhaust duct system often employs filtration to stop or slow the flow of unwanted material that is transported, through the use of a baghouse.

- In the laboratory, a Büchner funnel is often used, with a filter paper serving as the porous barrier.
- Air filters are commonly used to remove airborne particulate matter in building ventilation systems, combustion engines, and industrial processes.
- Oil filter in automobiles, often as a canister or cartridge.
- Aquarium filter.
- Straining water from food with a colander.

Adsorption filtration removes contaminants by adsorption of the contaminant by the filter medium. This requires intimate contact between the filter medium and the filtrate, and takes time for diffusion to bring the contaminant into direct contact with the medium while passing through it, referred to as dwell time. Slower flow also reduces pressure drop across the filter. Applications include:

- Carbon dioxide removal from breathing gas in rebreathers and life-support systems using scrubber filters,
- Activated carbon filters to remove volatile hydrocarbons, odours, and other contaminants from recirculated breathing gas in closed habitats.

Combined applications include:

- Compressed breathing air production, where the air passes through a particulate filter before entering the compressor, which removes particles likely to damage the compressor, followed by droplet separation after post-compression cooling and final product adsorption filtration to remove gaseous hydrocarbons contaminants and excessive water vapour. In some cases prefilters using adsorption media are used to control carbon dioxide levels, pressure swing adsorption may be used to increase oxygen fraction, and where the risk of carbon monoxide contamination exists, hopcalite catalytic converters may be included in the filtration media of the product. All these processes are broadly referred to as aspects of the filtration of the product.
- Potable water treatment using biofilm filtration in slow sand filters.
- Wastewater treatment using biofilm filtration using trickling filters.

CHAPTER

Centrifugation

4

Centrifugation is a mechanical process which involves the use of the centrifugal force to separate particles from a solution according to their size, shape, density, medium viscosity and rotor speed. The denser components of the mixture migrate away from the axis of the centrifuge, while the less dense components of the mixture migrate towards the axis. Chemists and biologists may increase the effective gravitational force of the test tube so that the precipitate (pellet) will travel quickly and fully to the bottom of the tube. The remaining liquid that lies above the precipitate is called a supernatant or supernate.

There is a correlation between the size and density of a particle and the rate that the particle separates from a heterogeneous mixture, when the only force applied is that of gravity. The larger the size and the larger the density of the particles, the faster they separate from the mixture. By applying a larger effective gravitational force to the mixture, like a centrifuge does, the separation of the particles is accelerated. This is ideal in industrial and lab settings because particles that would naturally separate over a long period of time can be separated in much less time.

The rate of centrifugation is specified by the angular velocity usually expressed as revolutions per minute (RPM), or acceleration expressed as g. The conversion factor between RPM and g depends on the radius of the centrifuge rotor. The particles' settling velocity in centrifugation is a function of their size and shape, centrifugal acceleration, the volume fraction of solids present, the density difference between the particle and the liquid, and the viscosity. The most common application is the separation of solid from highly concentrated suspensions, which is used in the treatment of sewage sludges for dewatering where less consistent sediment is produced.

The centrifugation method has a wide variety of industrial and laboratorial applications; not only is this process used to separate two miscible substances, but also to analyze the hydrodynamic properties of macromolecules. It is one of the most important and commonly used research methods in biochemistry, cell and molecular biology. In the chemical and food industries, special centrifuges can process a continuous stream of particle turning into separated liquid like plasma. Centrifugation is also the most common method used for uranium enrichment, relying on the slight mass difference between atoms of U-238 and U-235 in uranium hexafluoride gas.

4.1 HISTORY

By 1923 Theodor Svedberg and his student H. Rinde had successfully analyzed large-grained sols in terms of their gravitational sedimentation. Sols consist of a substance evenly distributed in another substance, also known as a colloid. However, smaller grained sols, such as those containing gold, could not be analyzed. To investigate this problem Svedberg developed an analytical centrifuge, equipped with a photographic absorption system, which would exert a much greater centrifugal effect. In addition, he developed the theory necessary to measure molecular weight. During this time, Svedberg's attention shifted from gold to proteins.

By 1900, it had been generally accepted that proteins were composed of amino acids; however, whether proteins were colloids or macromolecules was still under debate. One protein being investigated at the time was hemoglobin. It was determined to have 712 carbon, 1,130 hydrogen, 243 oxygen, two sulfur atoms, and at least one iron atom. This gave hemoglobin a resulting weight of approximately 16,000 dalton (Da) but it was uncertain whether this value was a multiple of one or four (dependent upon the number of iron atoms present).

Through a series of experiments utilizing the sedimentation equilibrium technique, two important observations were made: hemoglobin has a molecular weight of 68,000 Da, suggesting that there are four iron atoms present rather than one, and that, no matter where the hemoglobin was isolated from, it had exactly the same molecular weight. How something of such a large molecular mass could be consistently found, regardless of where it was sampled from in the body, was unprecedented and favored the idea that proteins are macromolecules rather than colloids. In order to investigate this phenomenon, a centrifuge with even higher speeds was needed, and thus the ultracentrifuge was created to apply the theory of sedimentation-diffusion. The same molecular mass was determined, and the presence of a spreading boundary suggested that it was a single compact particle. Further application of centrifugation showed that under different conditions the large homogeneous particles could be broken down into discrete subunits. The development of centrifugation was a great advance in experimental protein science.

Linderstorm-Lang, in 1937, discovered that density gradient tubes could be used for density measurements. He discovered this when working with potato yellow-dwarf virus. This method was also used in Meselson and Stahl's famous experiment in which they proved that DNA replication is semi-conservative by using different isotopes of nitrogen. They used density gradient centrifugation to determine which isotope or isotopes of nitrogen were present in the DNA after cycles of replication.

4.2 MATHEMATICAL FORMULA

In a liquid suspension, many particles or cells will gradually fall to the bottom of the container due to gravity; however, the amount of time taken for such separations is not feasible. Other particles, which are very small, can not be isolated at all in solution until they are exposed to a high centrifugal force. As the suspension is rotated at a certain speed or revolutions per minute (RPM), the centrifugal force allows the particles to travel radially away from the rotation axis. The general formula for calculating the revolutions per minute (RPM) of a centrifuge is:

$$RPM = \sqrt{\frac{g}{r}}$$

where g represents the relative centrifugal force (RCF) and r the radius from the center of the rotor to a point in the sample.

However, depending on the centrifuge model used, the respective angle of the rotor and the radius may vary, thus the formula gets modified. For example, the Sorvall #SS-34 rotor has a maximum radius of 10.8 cm, so the formula becomes $RPM = 299\sqrt{\frac{g}{r}}$, which can further simplify to $RPM = 91\sqrt{g}$.

When compared to gravity, the particle force is called the 'Relative Centrifugal Force' (RCF). It is the perpendicular force exerted on the contents of the rotor as a result of the rotation, always relative to the gravity of the Earth, which measures the strength of rotors of different types and sizes. For instance, the RCF of 1000 x g means that the centrifugal force is 1000 times stronger than the Earth's gravitational force. RCF is dependent on the speed of rotation in rpm and the distance of the particles from the center of rotation. The most common formula used for calculating RCF is:

$$RCF = 1.118 \times 10^{-5} \times r \times (rpm)^2$$

where 1.118×10^{-5} is a constant; r is the radius, expressed in centimetres, between the axis of rotation and a point in the sample; and rpm is the speed in revolutions per minute.

4.3 CENTRIFUGATION IN BIOLOGICAL RESEARCH

4.3.1 Microcentrifuges

Microcentrifuges are specially designed table-top models with light, small-volume rotors capable of very fast acceleration up to approximately 17,000 rpm. They are lightweight devices which are primarily used for short-time centrifugation of samples up to around 0.2–2.0 mL. However, due to their small scale, they are readily transportable and, if necessary, can be operated in a cold room. They can be refrigerated or not. The microcentrifuge is normally used in research laboratories where small samples of biological molecules, cells, or nuclei are required to be subjected to high RCF for relatively short time intervals. Microcentrifuges designed for high-speed operation can reach up to 35,000 rpm, giving RCF up to 30000×g, and are called high-speed microcentrifuges.

4.3.2 Low-Speed Centrifuges

Low-speed centrifuges are used to harvest chemical precipitates, intact cells (animal, plant and some microorganisms), nuclei, chloroplasts, large mitochondria and the larger plasma-membrane fragments. Density gradients for purifying cells are also run in these centrifuges. Swinging-bucket rotors tend to be used very widely because of the huge flexibility of sample size through the use of adaptors. These machines have maximum rotor speeds of less than 10 000 rpm and vary from small, bench-top to large, floor-standing centrifuges.

4.3.3 High-Speed Centrifuges

High-speed centrifuges are typically used to harvest microorganisms, viruses, mitochondria, lysosomes, peroxisomes and intact tubular Golgi membranes. The majority of the simple pelleting tasks are carried out in fixed angle rotors. Some density-gradient work for purifying cells and organelles can be carried out in swinging-bucket rotors, or in the case of Percoll gradients in fixed-angle rotors. High-speed or superspeed centrifuges can handle larger sample volumes, from a few tens of millilitres to several litres. Additionally, larger centrifuges can also reach higher angular velocities (around 30,000 rpm). The rotors may come with different adapters to hold various sizes of test tubes, bottles, or microtiter plates.

4.3.4 Ultracentrifugations

Ultracentrifugation makes use of high centrifugal force for studying properties of biological particles at exceptionally high speeds. Current ultracentrifuges can spin to as much as 150,000 rpm (equivalent to 1,000,000 x g). They are used to harvest all membrane vesicles derived from the plasma membrane, endoplasmic reticulum (ER) and Golgi membrane, endosomes, ribosomes, ribosomal subunits, plasmids, DNA, RNA and proteins in fixed-angle rotors. Compared to microcentrifuges or high-speed centrifuges, ultracentrifuges can isolate much smaller particles and, additionally, whilst microcentrifuges and supercentrifuges separate particles in batches (limited volumes of samples must be handled manually in test tubes or bottles), ultracentrifuges can separate molecules in batch or continuous flow systems.

Ultracentrifugation is employed for separation of macromolecules/ligand binding kinetic studies, separation of various lipoprotein fractions from plasma and deprotonisation of physiological fluids for amino acid analysis.

They are the most commonly used centrifuge for the density-gradient purification of all particles except cells, and, whilst swinging buckets have been traditionally used for this purpose, fixed-angle rotors and vertical rotors are also used, particularly for self-generated gradients and can improve the efficiency of separation greatly. There are two kinds of ultracentrifuges: the analytical and the preparative.

Analytical Ultracentrifugation

Analytical ultracentrifugation (AUC) can be used for determination of the properties of macromolecules such as shape, mass, composition, and conformation. It is a commonly used biomolecular analysis technique used to evaluate sample purity, to characterize the assembly and disassembly mechanisms of biomolecular complexes, to determine subunit stoichiometries, to identify and characterize macromolecular conformational changes, and to calculate equilibrium constants and thermodynamic parameters for self-associating and hetero-associating systems. Analytical ultracentrifuges incorporate a scanning visible/ultraviolet light-based optical detection system for real-time monitoring of the sample's progress during a spin.

Samples are centrifuged with a high-density solution such as sucrose, caesium chloride, or iodixanol. The high-density solution may be at a uniform concentration throughout the test tube ("cushion") or a varying concentration ("gradient"). Molecular properties can be modeled through sedimentation velocity analysis or sedimentation equilibrium analysis. During the run, the particle or molecules will migrate through the test tube at different speeds depending on their physical properties and the properties of the solution, and eventually form a pellet at the bottom of the tube, or bands at various heights.

Preparative Ultracentrifugation

Preparative ultracentrifuges are often used for separating particles according to their densities, isolating and/or harvesting denser particles for collection in the pellet, and clarifying suspensions containing particles. Sometimes researchers also use preparative ultracentrifuges if they need the flexibility to change the type of rotor in the instrument. Preparative ultracentrifuges can be equipped with a wide range of different rotor types, which can spin samples of different numbers, at different angles, and at different speeds.

4.3.5 Fractionation Process

In biological research, cell fractionation typically includes the isolation of cellular components while retaining the individual roles of each component. Generally, the cell sample is stored in a suspension which is:

- Buffered—neutral pH, preventing damage to the structure of proteins including enzymes (which could affect ionic bonds).
- Isotonic (of equal water potential)—this prevents water gain or loss by the organelles.
- Cool—reducing the overall activity of enzyme released later in the procedure.

Centrifugation is the first step in most fractionations. Through low-speed centrifugation, cell debris may be removed, leaving a supernatant preserving the contents of the cell. Repeated centrifugation at progressively higher speeds will fractionate homogenates of cells into their components. In general, the smaller the subcellular component, the greater is the centrifugal force required to sediment it. The soluble fraction of any lysate can then be further separated into its constituents using a variety of methods.

4.3.6 Differential Centrifugation

Differential centrifugation is the simplest method of fractionation by centrifugation, commonly used to separate organelles and membranes found in cells. Organelles generally differ from each other in density and in size, making the use of differential centrifugation, and centrifugation in general, possible. The organelles can then be identified by testing for indicators that are unique to the specific organelles. The most widely used application of this technique is to produce crude subcellular fractions from a tissue homogenate such as that from rat liver. Particles of different densities or sizes in a suspension are sedimented at different rates, with the larger and denser particles sedimenting faster. These sedimentation rates can be increased by using centrifugal force.

A suspension of cells is subjected to a series of increasing centrifugal force cycles to produce a series of pellets comprising cells with a declining sedimentation rate. Homogenate includes nuclei, mitochondria, lysosomes, peroxisomes, plasma membrane sheets and a broad range of vesicles derived from a number of intracellular membrane compartments and also from the plasma membrane, typically in a buffered medium.

4.3.7 Density Gradient Centrifugation

Density gradient centrifugation is known to be one of the most efficient methods for separating suspended particles, and is used both as a separation technique and as a method for measuring the density of particles or molecules in a mixture. It is used to separate particles on the basis of size, shape, and density by using a medium of graded densities. During a relatively short or slow centrifugation, the particles are separated by size, with larger particles sedimenting farther than smaller ones. Over a long or fast centrifugation, particles travel to locations in the gradient where the density of the medium is the same as that of the particle density; $(\rho p - \rho m) \rightarrow 0$. Therefore, a small, dense particle initially sediments less readily than a large, low density particle. The large particles reach their equilibrium density position early, while the small particles slowly migrate across the large particle zone and ultimately take up an equilibrium position deeper into the gradient.

A tube, after being centrifuged by this method, has particles in order of density based on height. The object or particle of interest will reside in the position within the tube corresponding to its density. Nevertheless, some non-ideal sedimentations are still possible when using this method. The first potential issue is the unwanted aggregation of particles, but this can occur in any centrifugation. The second possibility occurs when droplets of solution that contain particles sediment. This is more likely to occur when working with a solution that has a layer of suspension floating on a dense liquid, which in fact have little to no density gradient.

4.4 OTHER APPLICATIONS

A centrifuge can be used to isolate small quantities of solids retained in suspension from liquids, such as in the separation of chalk powder from water. In biological research, it can be used in the purification of mammalian cells, fractionation of subcellular organelles, fractionation of membrane vesicles, fractionation of macromolecules and macromolecular complexes, etc. Centrifugation is used in many different ways in the food industry. For example, in the dairy industry, it is typically used in the clarification and skimming of milk, extraction of cream, production and recovery of casein, cheese production, removing bacterial contaminants, etc. This processing technique is also used in the production of beverages, juices, coffee, tea, beer, wine, soy milk, oil and fat processing/ recovery, cocoa butter, sugar production, etc. It is also used in the clarification and stabilization of wine.

In forensic and research laboratories, it can be used in the separation of urine and blood components. It also aids in separation of proteins using purification techniques such as salting out, e.g. ammonium sulfate precipitation. Centrifugation is also an important technique in waste treatment, being one of the most common processes used for sludge dewatering. This process also plays a role in cyclonic separation, where particles are separated from an air-flow without the use of filters. In a cyclone collector, air moves in a helical path. Particles with high inertia are separated by the centrifugal force whilst smaller particles continue with the air-flow.

Centrifuges have also been used to a small degree to isolate lighter-than-water compounds, such as oil. In such situations, the aqueous discharge is obtained at the opposite outlet from which solids with a specific gravity greater than one are the target substances for separation.

CHAPTER

Adsorption

5

Adsorption is the adhesion of atoms, ions or molecules from a gas, liquid or dissolved solid to a surface. This process creates a film of the adsorbate on the surface of the adsorbent.

This process differs from absorption, in which a fluid (the absorbate) is dissolved by or permeates a liquid or solid (the absorbent). While adsorption does often precede absorption, which involves the transfer of the absorbate into the volume of the absorbent material, alternatively, adsorption is distinctly a surface phenomenon, wherein the adsorbate does not penetrate through the material surface and into the bulk of the adsorbent. The term sorption encompasses both adsorption and absorption, and desorption is the reverse of sorption.

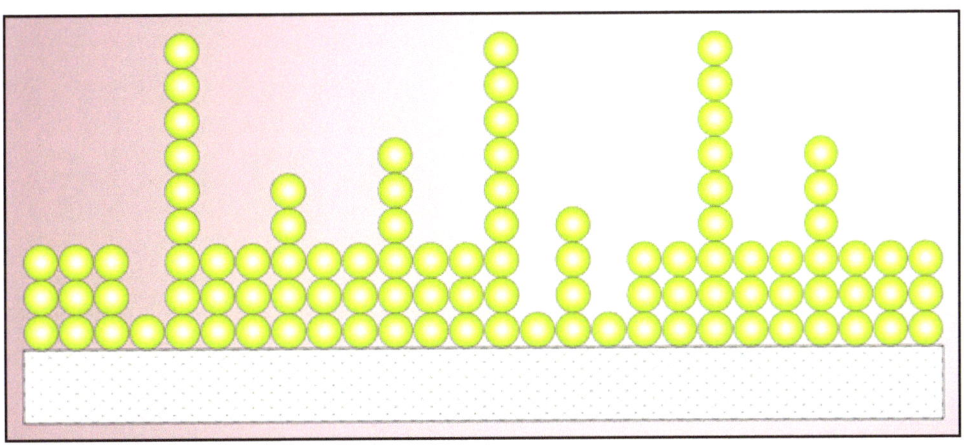

Fig: Brunauer, Emmett and Teller's Model of Multilayer Adsorption is a Random Distribution of Molecules on the Material Surface

Like surface tension, adsorption is a consequence of surface energy. In a bulk material, all the bonding requirements (be they ionic, covalent or metallic) of the constituent atoms of the material are fulfilled by other atoms in the material. However, atoms on the surface of the adsorbent are not wholly surrounded by other adsorbent atoms and therefore can attract adsorbates. The exact nature of the bonding depends on the details of the species involved, but the adsorption process is generally classified as physisorption (characteristic of weak van der

Waals forces) or chemisorption (characteristic of covalent bonding). It may also occur due to electrostatic attraction. The nature of the adsorption can affect the structure of the adsorbed species. For example, polymer physisorption from solution can result in squashed structures on a surface.

Adsorption is present in many natural, physical, biological and chemical systems and is widely used in industrial applications such as heterogeneous catalysts, activated charcoal, capturing and using waste heat to provide cold water for air conditioning and other process requirements (adsorption chillers), synthetic resins, increasing storage capacity of carbide-derived carbons and water purification. Adsorption, ion exchange and chromatography are sorption processes in which certain adsorbates are selectively transferred from the fluid phase to the surface of insoluble, rigid particles suspended in a vessel or packed in a column. Pharmaceutical industry applications, which use adsorption as a means to prolong neurological exposure to specific drugs or parts thereof, are lesser known.

5.1 ADSORBENTS

5.1.1 Characteristics and General Requirements

Fig: Activated Carbon is Used as an Adsorbent

Adsorbents are used usually in the form of spherical pellets, rods, moldings, or monoliths with a hydrodynamic radius between 0.25 and 5 mm. They must have high abrasion resistance, high thermal stability and small pore diameters, which results in higher exposed surface area and hence high capacity for adsorption. The adsorbents must also have a distinct pore structure that enables fast transport of the gaseous vapors. Most industrial adsorbents fall into one of three classes:

- **Oxygen-Containing Compounds**—Are typically hydrophilic and polar, including materials such as silica gel, limestone (calcium carbonate) and zeolites.
- **Carbon-Based Compounds**—Are typically hydrophobic and non-polar, including materials such as activated carbon and graphite.
- **Polymer-Based Compounds**—Are polar or non-polar, depending on the functional groups in the polymer matrix.

5.1.2 Silica Gel

Fig: Silica gel adsorber for NO_2, Fixed Nitrogen Research Laboratory, ca.1930s

Silica gel is a chemically inert, non-toxic, polar and dimensionally stable (< 400 °C or 750 °F) amorphous form of SiO_2. It is prepared by the reaction between sodium silicate and acetic acid, which is followed by a series of after-treatment processes such as aging, pickling, etc. These after-treatment methods results in various pore size distributions. Silica is used for drying of process air (e.g. oxygen, natural gas) and adsorption of heavy (polar) hydrocarbons from natural gas.

5.1.3 Zeolites

Zeolites are natural or synthetic crystalline aluminosilicates, which have a repeating pore network and release water at high temperature. Zeolites are polar in nature. They are manufactured by hydrothermal synthesis of sodium aluminosilicate or another silica source in an autoclave followed by ion exchange with certain cations (Na^+, Li^+, Ca^{2+}, K^+, NH_4^+). The channel diameter of zeolite cages usually ranges from 2 to 9 Å. The ion exchange process is followed by drying of the crystals, which can be pelletized with a binder to form macroporous pellets.

Zeolites are applied in drying of process air, CO_2 removal from natural gas, CO removal from reforming gas, air separation, catalytic cracking, and catalytic synthesis and reforming. Non-polar (siliceous) zeolites are synthesized from aluminum-free silica sources or by dealumination of aluminum-containing zeolites. The dealumination process is done by treating the zeolite with steam at elevated temperatures, typically greater than 500 °C (930 °F). This high temperature heat treatment breaks the aluminum-oxygen bonds and the aluminum atom is expelled from the zeolite framework.

5.1.4 Activated Carbon

The term "adsorption" itself was coined by Heinrich Kayser in 1881 in the context of uptake of gases by carbons. Activated carbon is a highly porous, amorphous solid consisting of microcrystallites with a graphite lattice, usually prepared in small pellets or a powder. It is non-polar and cheap. One of its main drawbacks is that it reacts with oxygen at moderate temperatures (over 300 °C).

Activated carbon can be manufactured from carbonaceous material, including coal (bituminous, subbituminous, and lignite), peat, wood, or nutshells (e.g., coconut). The manufacturing process consists of two phases, carbonization and activation. The carbonization process includes drying and then heating to separate by-products, including tars and other hydrocarbons from the raw material, as well as to drive off any gases generated. The process is completed by heating the material over 400 °C (750 °F) in an oxygen-free atmosphere that cannot support combustion. The carbonized particles are then "activated" by exposing them to an oxidizing agent, usually steam or carbon dioxide at high temperature. This agent burns off the pore blocking structures created during the carbonization phase and so, they develop a porous, three-dimensional graphite lattice structure. The size of the pores developed during activation is a function of the time that they spend in this stage. Longer exposure times result in larger pore sizes. The most popular aqueous phase carbons are bituminous based because of their hardness, abrasion resistance, pore size distribution, and low cost, but their effectiveness needs to be tested in each application to determine the optimal product.

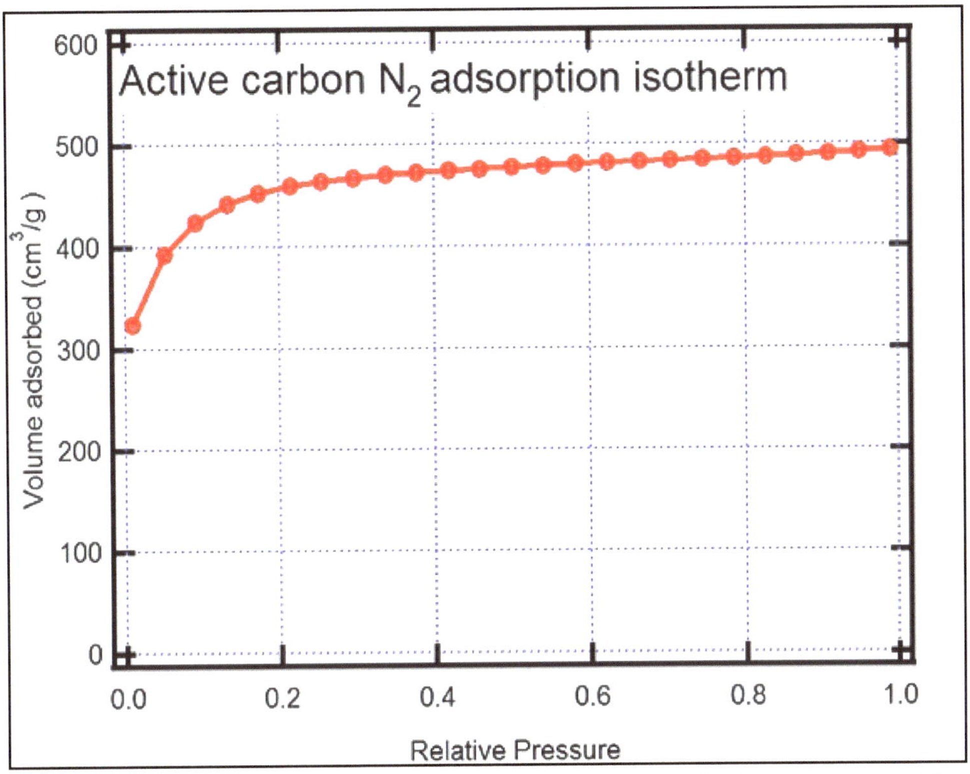

Fig: Activated Carbon Nitrogen Isotherm Showing a Marked Microporous Type I Behavior

Activated carbon is used for adsorption of organic substances and non-polar adsorbates and it is also usually used for waste gas (and waste water) treatment. It is the most widely used adsorbent since most of its chemical (e.g. surface groups) and physical properties (e.g. pore size distribution and surface area) can be tuned according to what is needed. Its usefulness also derives from its large micropore (and sometimes mesopore) volume and the resulting high surface area. Recent research works reported activated carbon as an effective agent to adsorb cationic species of toxic metals from multi-pollutant systems and also proposed possible adsorption mechanisms with supporting evidences.

5.2 WATER ADSORPTION

The adsorption of water at surfaces is of broad importance in chemical engineering, materials science and catalysis. Also termed surface hydration, the presence of physically or chemically adsorbed water at the surfaces of solids plays an important role in governing interface properties, chemical reaction pathways and catalytic performance in a wide range of systems. In the case of physically adsorbed water, surface hydration can be eliminated simply through drying at conditions of temperature and pressure allowing full vaporization of water. For chemically adsorbed water, hydration may be in the form of either dissociative adsorption, where H_2O molecules are dissociated into surface adsorbed -H and -OH, or molecular adsorption (associative adsorption) where individual water molecules remain intact

5.3 ADSORPTION SOLAR HEATING AND STORAGE

The low cost ($200/ton) and high cycle rate (2,000 ×) of synthetic zeolites such as Linde 13X with water adsorbate has garnered much academic and commercial interest recently for use for thermal energy storage (TES), specifically of low-grade solar and waste heat. Several pilot projects have been funded in the EU from 2000 to the present (2020). The basic concept is to store solar thermal energy as chemical latent energy in the zeolite. Typically, hot dry air from flat plate solar collectors is made to flow through a bed of zeolite such that any water adsorbate present is driven off. Storage can be diurnal, weekly, monthly, or even seasonal depending on the volume of the zeolite and the area of the solar thermal panels. When heat is called for during the night, or sunless hours, or winter, humidified air flows through the zeolite. As the humidity is adsorbed by the zeolite, heat is released to the air and subsequently to the building space. This form of TES, with specific use of zeolites, was first taught by John Guerra in 1978.

5.4 CARBON CAPTURE AND STORAGE

Typical adsorbents proposed for carbon capture and storage are zeolites and MOFs. The customization of adsorbents makes them a potentially attractive alternative to absorption. Because adsorbents can be regenerated by temperature or pressure swing, this step can be less energy intensive than absorption regeneration methods. Major problems that are present with adsorption cost in carbon capture are: regenerating the adsorbent, mass ratio, solvent/MOF, cost of adsorbent, production of the adsorbent, lifetime of adsorbent. In sorption enhanced water gas shift (SEWGS) technology a pre-combustion carbon capture process, based on solid adsorption, is combined with the water gas shift reaction (WGS) in order to produce a high pressure hydrogen stream. The CO_2 stream produced can be stored or used for other industrial processes.

5.5 PROTEIN ADSORPTION

Adsorption is the accumulation and adhesion of molecules, atoms, ions, or larger particles to a surface, but without surface penetration occurring. The adsorption of larger biomolecules such as proteins is of high physiological relevance, and as such they adsorb with different mechanisms than their molecular or atomic analogs. Some of the major driving forces behind protein adsorption include:

surface energy, intermolecular forces, hydrophobicity, and ionic or electrostatic interaction. By knowing how these factors affect protein adsorption, they can then be manipulated by machining, alloying, and other engineering techniques to select for the most optimal performance in biomedical or physiological applications.

5.5.1 Relevance

Many medical devices and products come into contact with the internal surfaces of the body, such as surgical tools and implants. When a non-native material enters the body, the first step of the immune response takes place and host extracellular matrix and plasma proteins aggregate to the material in attempts to contain, neutralize, or wall-off the injurious agent. These proteins can facilitate the attachment of various cell types such as osteoblasts and fibroblasts that can encourage tissue repair. Taking this a step further, implantable devices can be coated with a bioactive material to encourage adsorption of specific proteins, fibrous capsule formation, and wound healing. This would reduce the risk of implant rejection and accelerate recovery by selecting for the necessary proteins and cells necessary for endothelialization. After the formation of the endothelium, the body will no longer be exposed to the foreign material, and will stop the immune response.

Proteins such as collagen or fibrin often serve as scaffolds for cell adhesion and cell growth. This is an integral part to the structural integrity of cell sheets and their differentiation into more complex tissue and organ structures. The adhesion properties of proteins to non-biological surfaces greatly influences whether or not cells can indirectly attach to them via scaffolds. An implant like a hip-stem replacement necessitates integration with the host tissues, and protein adsorption facilitates this integration.

Surgical tools can be designed to be sterilized more easily so that proteins do not remain adsorbed to a surface, risking cross-contamination. Some diseases such as Creutzfeldt–Jakob disease and kuru (both related to mad cow disease) are caused by the transmission of prions, which are errant or improperly folded forms of a normally native protein. Surgical tools contaminated with prions require a special method of sterilization to completely eradicate all trace elements of the misfolded protein, as they are resistant to many of the normally used cleansing methods.

However, in some cases, protein adsorption to biomaterials can be an extremely unfavorable event. The adhesion of clotting factors may induce thrombosis, which may lead to stroke or other blockages. Some devices are intended to interact with the internal body environment such as sensors or drug-delivery vehicles, and protein adsorption would hinder their effectiveness.

5.5.2 Fundamentals of Protein Adsorption

Proteins are biomolecules that are composed of amino acid subunits. Each amino acid has a side chain that gains or loses charge depending on the pH of the surrounding environment, as well as its own individual polar/nonpolar qualities.

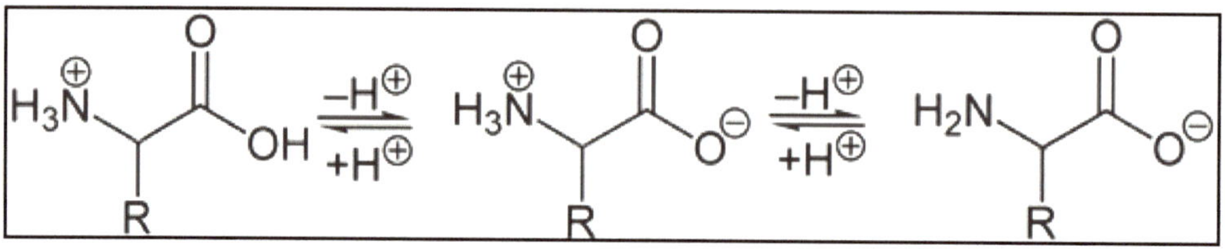

Fig: Amino Acid Titration

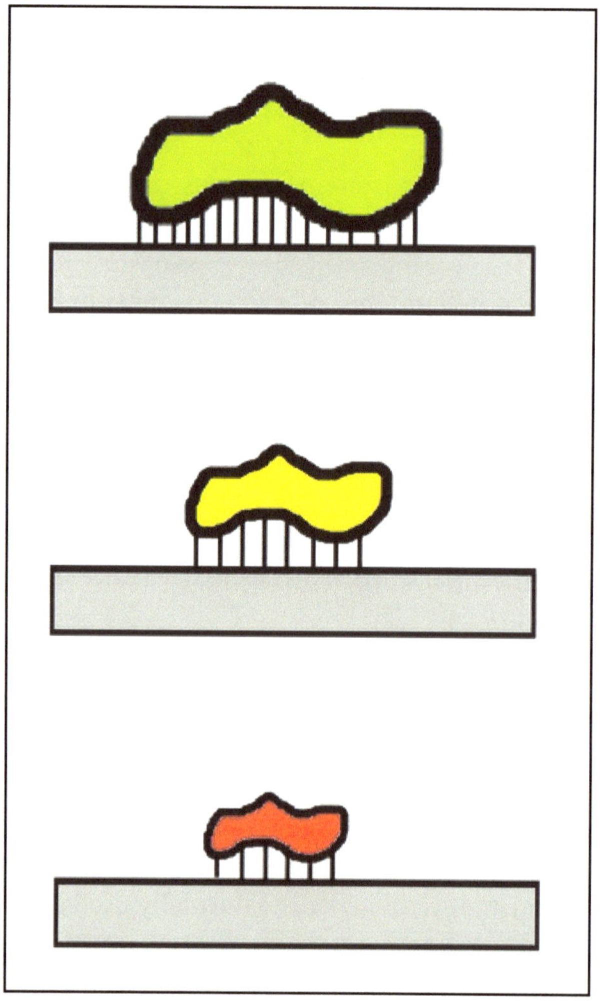

Fig: The Effect of Protein Size on Interaction with a Surface. Notice that the Larger Protein Composed of More Amino Acids is Capable of Making more Interactions

Charged regions can greatly contribute to how that protein interacts with other molecules and surfaces, as well as its own tertiary structure (protein folding). As a result of their hydrophilicity, charged amino acids tend to be located on the outside of proteins, where they are able to interact with surfaces. It is the unique combination of amino acids that gives a protein its properties. In terms of surface chemistry, protein adsorption is a critical phenomenon that describes the aggregation of these molecules on the exterior of a material. The tendency for proteins to remain attached to a surface depends largely on the material properties such as surface energy, texture, and relative charge distribution. Larger proteins are more likely to adsorb and remain attached to a surface due to the higher number of contact sites between amino acids and the surface.

Energy of Protein Adsorption

The fundamental idea behind spontaneous protein adsorption is that adsorption occurs when more energy is released than gained according to Gibbs law of free energy.

This is seen in the equation:

$$\Delta_{ads}G = \Delta_{ads}H - T\Delta_{ads}S < 0$$

Vroman Effect

Proteins and other molecules are constantly in competition with one another over binding sites on a surface. The Vroman Effect, developed by Leo Vroman, postulates that small and abundant molecules will be the first to coat a surface. However, over time, molecules with higher affinity for that particular surface will replace them. This is often seen in materials that contact the blood where fibrinogen will bind to the surface first and over time will be replaced by kininogen.

Rate of Adsorption

In order for proteins to adsorb, they must first come into contact with the surface through one or more of these major transport mechanisms: diffusion, thermal convection, bulk flow, or a combination thereof. When considering the transport of proteins, it is clear how concentration gradients, temperature, protein size and flow velocity will influence the arrival of proteins to a solid surface. Under conditions of low flow and minimal temperature gradients, the adsorption rate can be modeled after the diffusion rate equation.

Diffusion Rate Equation

$$\frac{dn}{dt} = C_o\left(\frac{D}{\pi t}\right)^{1/2}$$

A higher bulk concentration and/or higher diffusion coefficient (inversely proportional to molecular size) results in a larger number of molecules arriving at the surface. The consequential protein surface interactions result in high local concentrations of adsorbed protein, reaching concentrations of up to 1000 times higher than in the bulk solution. However, the body is much more complex, containing flow and convective diffusion, and these must be considered in the rate of protein adsorption.

Flow in a Thin Channel

$$\frac{\partial C}{\partial t} + V(y)\frac{\partial C}{\partial x} = D\frac{\partial^2 C}{\partial y^2}$$

and

$$V(y) = \gamma y(1 - \frac{y}{b})$$

5.5.3 Forces and Interactions Influencing Protein Adsorption

The four fundamental classes of forces and interaction in protein adsorption are: 1) ionic or electrostatic interaction, 2) hydrogen bonding, 3) hydrophobic interaction (largely entropically driven), and 4) interactions of charge-transfer or particle electron donor/acceptor type.

Ionic or Electrostatic Interactions

The charge of proteins is determined by the pKa of its amino acid side chains, and the terminal amino acid and carboxylic acid. Proteins with isoelectric point (pI) above physiological conditions have a positive charge and proteins with pI below physiological conditions have a negative charge. The net charge of the protein, determined by the sum charge of its constituents, results in electrophoretic migration in a physiologic electric field. These effects are short-range because of the high di-electric constant of water, however, once the protein is close to a charged surface, electrostatic coupling becomes the dominant force.

Hydrogen Bonding

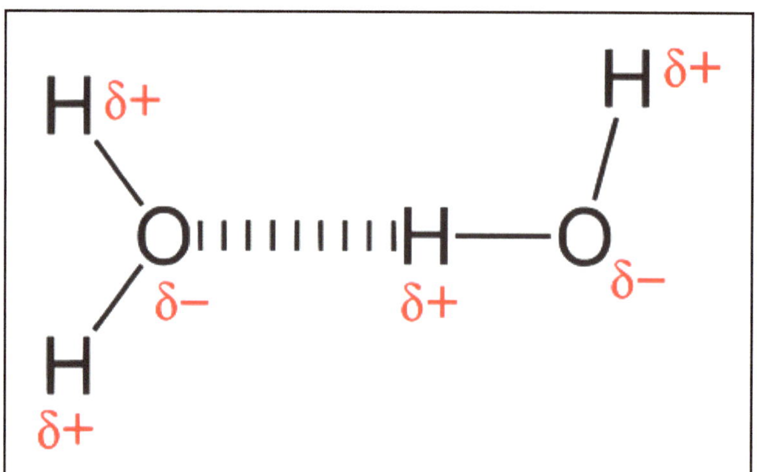

Fig: Illustration of Two Water Molecules Interacting to form a Hydrogen Bond

Water has as much propensity to form hydrogen bonds as any group in a polypeptide. During a folding and association process, peptide and amino acid groups exchange hydrogen bonds with water. Thus, hydrogen bonding does not have a strong stabilizing effect on protein adsorption in an aqueous medium.

Hydrophobic Interactions

Hydrophobic interactions are essentially entropic interactions basically due to order/disorder phenomena in an aqueous medium. The free energy associated with minimizing interfacial areas is responsible for minimizing the surface area of water droplets and air bubbles in water.

This same principle is the reason that hydrophobic amino acid side chains are oriented away from water, minimizing their interaction with water. The hydrophilic groups on the outside of the molecule result in protein water solubility.

Characterizing this phenomenon can be done by treating these hydrophobic relationships with interfacial free energy concepts. Accordingly, one can think of the driving force of these interactions as the minimization of total interfacial free energy, i.e. minimization of surface area.

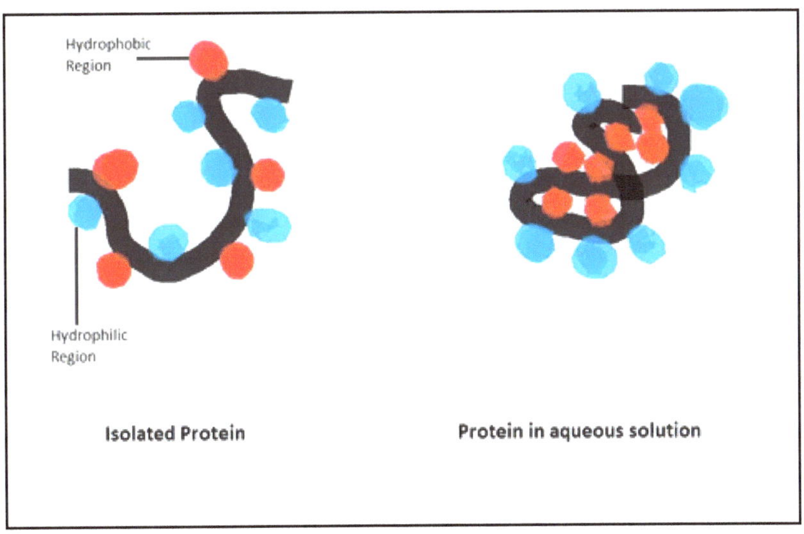

Fig: Illustration of how Protein Changes Shape to Allow Polar Regions (Blue) to Interact with Water while Non-Polar Hydrophobic Regions (Red) do not Interact with the Water

Charge-Transfer Interactions

Charge-transfer interactions are also important in protein stabilization and surface interaction. In general donor-acceptor processes, one can think of excess electron density being present which can be donated to an electrophilic species. In aqueous media, these solute interactions are primarily due to pi orbital electron effects.

5.5.4 Other Factors Influencing Protein Adsorption

Temperature

Temperature has an effect on both, the equilibrium state and kinetics of protein adsorption. The amount of protein adsorbed at high temperature is usually higher than that at room temperature. Temperature variation causes conformational changes in protein influencing adsorption. These conformational rearrangements in proteins results in an entropy gain which acts as a major driving force for protein adsorption.

The temperature effect on protein adsorption can be seen in food manufacturing processes, especially liquid foods such as, milk which causes severe fouling on the wall surfaces of equipment where thermal treatment is carried out.

Ionic Strength

Ionic strength determines the Debye length that correlates with the damping distance of the electric potential of a fixed charge in an electrolyte. So, higher the ionic strength the shorter are electrostatic interactions between charged entities. As a result, the adsorption of charged proteins to oppositely charged substrates is hindered whereas the adsorption to like charged substrates is enhanced, thereby influencing adsorption kinetics. Also, high ionic strength increases the tendency of proteins to aggregate.

Multi-Protein System

When a surface is exposed to a multi-protein solution, adsorption of certain protein molecules are favored over the others. Protein molecules approaching the surface compete for binding sites. In multi-protein system attraction between molecules can occur, whereas in single-protein solutions intermolecular repulsive interactions dominate. In addition, there is a time-dependent protein spreading, where protein molecules initially make contact with minimal binding sites on the surface. With the increase in protein's residence time on the surface, the protein may unfold for interaction with additional binding sites. This results in a time-dependent increase in the contact points between protein and surface. This further makes desorption less likely.

5.5.5 Experimental Approaches for Studying Protein Adsorption

Solution Depletion Technique

This technique measures a concentration change of proteins in bulk solution before and after adsorption, Δc_p. Any protein concentration change is attributed to the adsorbed layer, Γ_p.

$$\Gamma_p = \Delta c_p V/A_{tot}$$

where:

- V = total volume of protein solution
- A_{tot} = Total area available for adsorption

This method also requires a high surface area material such as particulate and beaded adsorbents.

Ellipsometry

Ellipsometry has been used widely for measuring protein adsorption kinetics as well as the structure of the adsorbed protein layer. It is an optical technique that measures the change of the polarization of light after reflection from a surface. This technique requires planar, reflecting surfaces, preferably quartz, silicon or silica, and a strong change in refractive index upon protein adsorption.

Atomic-Force Microscopy

Atomic-force microscopy (AFM) is a powerful microscopy technique used for studying samples at a nanoscale and is often used to image protein distribution on a surface. It consists of a cantilever with a tip to scan over the surface. It is a valuable tool for measuring protein-protein and protein-surface interaction. However, the limiting factor of many AFM studies is that imaging is often performed after drying the surface which might affect protein folding and the structure of the protein layer. Moreover, the cantilever tip can dislodge a protein or corrugate the protein layer.

Surface Plasmon Resonance

Surface plasmon resonance (SPR) has been widely used for measuring protein adsorption with high sensitivity. This technique is based on the excitation of surface plasmons, longitudinal electromagnetic waves originated at the interface between metals and dielectrics. The deposition on the conducting surface of molecules and thin layers within 200 nm modifies the dielectric properties of the system and thus the SPR response, signaling the presence of molecules on a metal surface.

Quartz Crystal Microbalance

Quartz crystal microbalance (QCM) is an acoustic sensor built around a disk shaped quartz crystal. It makes use of the converse piezoelectric effect. QCM, and extended versions such as QCM-D, has been widely used for protein adsorption studies, especially, real time monitoring of label-free protein adsorption. In addition to the adsorption studies, QCM-D also provides information regarding elastic moduli, viscosity, and conformational changes.

Optical Waveguide Lightmode Spectroscopy

Optical waveguide lightmode spectroscopy (OWLS) is a device that relies on a thin-film optical waveguide, enclosing a discrete number of guided electromagnetic waves. Guidance is achieved by means of a grating coupler. It is based on the measurements of effective refractive index of a thin-film layer above the waveguide. This technique works only on highly transparent surfaces. Other methods widely used for measuring the amount of protein adsorbed on surfaces include radio-labelling, Lowry assay, scanning angle reflectometry, total internal reflection fluorescence, bicinchoninic acid assay etc.

5.5.6 Protein Adsorption to Metals

Chemical Composition

Metallic bonding refers to the specific bonding between positive metal ions and surrounding valence electron clouds. This intermolecular force is relatively strong, and gives rise to the repeated crystalline orientation of atoms, also referred to as its lattice system. There are several types of common lattice formations, and each has its own unique packing density and atomic closeness. The negatively charged electron clouds of the metal ions will sterically hinder the adhesion of negatively charged protein regions due to charge repulsion, thus limiting the available binding sites of a protein to a metal surface.

The lattice formation can lead to connection with exposed potential metal-ion-dependent adhesion sites (MIDAS) which are binding sites for collagen and other proteins. The surface of the metal has different properties than the bulk since the normal crystalline repeating subunits is terminated at the surface. This leaves the surface atoms without a neighboring atom on one side, which inherently alters the electron distribution. This phenomenon also explains why the surface atoms have a higher energy than the bulk, often simply referred to as surface energy. This state of higher energy is unfavorable, and the surface atoms will try to reduce it by binding to available reactive molecules.

This is often accomplished by protein adsorption, where the surface atoms are reduced to a more advantageous energy state.

The internal environment of the body is often modeled to be an aqueous environment at 37 °C at pH 7.3 with plenty of dissolved oxygen, electrolytes, proteins, and cells. When exposed to oxygen for an extended period of time, many metals may become oxidized and increase their surface oxidation state by losing electrons. This new cationic state leaves the surface with a net positive charge, and a higher affinity for negatively charged protein side groups. Within the vast diversity of metals and metal alloys, many are susceptible to corrosion when implanted in the body. Elements that are more electronegative are corroded faster when exposed to an electrolyte-rich aqueous environment such as the human body. Both oxidation and corrosion will lower the free energy, thus affecting protein adsorption.

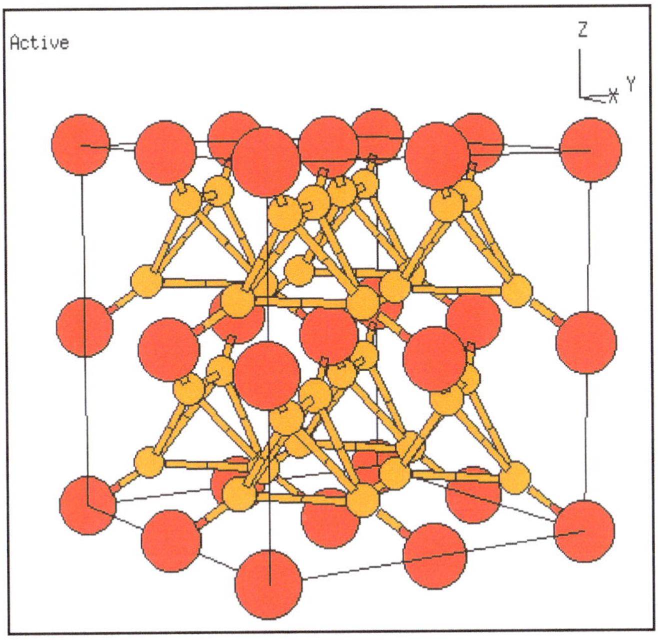

Fig: Notice in the Diagram of Fe4C that the Surface Atoms are Missing Neighboring Atoms

Effects of Topography

Surface roughness and texture has an undeniable influence on protein adsorption on all materials, but with the ubiquity of metal machining processes, it is useful to address how these impact protein behavior. The initial adsorption is important, as well as maintained adhesion and integrity. Research has shown that surface roughness can encourage the adhesion of scaffold proteins and osteoblasts, and results in an increase in surface mineralization. Surfaces with more topographical features and roughness will have more exposed surface area for proteins to interact with. In terms of biomedical engineering applications, micromachining techniques are often used to increase protein adhesion to implants in the hopes of shortening recovery time.

The technique of laserpatterning introduces grooves and surface roughness that will influence adhesion, migration and alignment. Grit-blasting, a method analogous to sand blasting, and chemical etching have proven to be successful surface roughening techniques that promote the long-term stability of titanium implants. The increase in stability is a direct result of the observed increase in extracellular matrix and collagen attachment, which results in increased osteoblast attachment and mineralization when compared to non-roughened surfaces. Adsorption is not always desirable, however. Machinery can be negatively affected by adsorption, particularly with Protein adsorption in the food industry.

5.5.7 Protein Adsorption to Polymers

Polymers are of great importance when considering protein adsorption in the biomedical arena. Polymers are composed of one or more types of "mers" bound together repeatedly, typically by directional covalent bonds. As the chain grows by the addition of mers, the chemical and physical properties of the material are dictated by the molecular structure of the monomer.

By carefully selecting the type or types of mers in a polymer and its manufacturing process, the chemical and physical properties of a polymer can be highly tailored to adsorb specific proteins and cells for a particular application.

Conformation Effects

Protein adsorption often results in significant conformational changes, which refers to changes in the secondary, tertiary, and quartary structures of proteins. In addition to adsorption rates and amounts, orientation and conformation are of critical importance. These conformational changes can affect protein interaction with ligands, substrates, and antigens which are dependent on the orientation of the binding site of interest. These conformational changes, as a result of protein adsorption, can also denature the protein and change its native properties.

Fig: Illustration of Protein (Green) Ligand (Red Star) Binding Site Alteration by the Conformational Change of the Protein as a Result of Surface (Blue) Adsorption. Note how the Ligand no Longer Fits into the Binding Site

Adsorption to Polymer Scaffolds

Tissue engineering is a relatively new field that utilizes a scaffolding as a platform upon which the desired cells proliferate. It is not clear what defines an ideal scaffold for a specific tissue type. The considerations are complex and protein adsorption only adds to the complexity. Although architecture, structural mechanics, and surface properties play a key role, understanding degradation and rate of protein adsorption are also key. In addition to the essentials of mechanics and geometry, a suitable scaffold construct will possess surface properties that are optimized for the attachment and migration of the cell types of particular interest.

Generally, it has been found that scaffolds that closely resemble the natural environments of the tissue being engineered are the most successful. As a result, much research has gone into investigating natural polymers that can be tailored, through processing methodology, toward specific design criteria. Chitosan is currently one of the most widely used polymers as it is very similar to naturally occurring glycosaminoglycan (GAGs) and it is degradable by human enzymes.

Chitosan

Chitosan is a linear polysaccharide containing linked chitin-derived residues and is widely studied as a biomaterial due to its high compatibility with numerous proteins in the body. Chitosan is cationic and thus electrostatically reacts with numerous proteoglycans, anionic GAGs, and other molecules possessing a negative charge. Since many cytokines and growth factors are linked to GAG, scaffolds with the chitosan-GAG complexes are able to retain these proteins secreted by the adhered cells. Another quality of chitosan that gives it good biomaterial potential is its high charge density in solutions. This allows chitosan to form ionic complexes with many water-soluble anionic polymers, expanding the range of proteins that are able to bind to it and thus expanding its possible uses.

Table: Structures, Target Tissues, and Application Cell Types of Chitosan-Based Scaffolds

Polymer	Scaffold Structure	Target Tissue	Application Cell Type
Chitosan	3D porous blocks	Bone	Osteoblast-like ROS
Chitosan-polyester	3D fiber meshes	Bone	Human MSC
Chitosan-alginate	Injectable gel	Bone	Osteoblast-like MG63
Chitosan-gelatin	3D porous cylinders	Cartilage	Chondrocytes
Chitosan-GP	Injectable gel	Cartilage	Chondrocytes
Chitosan-collagen	Porous membranes	Skin	Fibroblast and keratinocyte co-culture

5.5.8 Protein Adsorption at Fluid Interfaces

Due to their amphiphilic chemistry proteins are surface active and adsorb at fluid interfaces. In multiphase systems like emulsions or foams proteins adsorb at the oil-water or air-water interface, respectively, and reduce the interface tension, thereby increasing their stability. Proteins are pH-dependent polyampholytes that undergo structural rearrangements upon adsorption, wherefore their adsorption depends on the protein thermodynamic stability, the pH and ionic strength of the aqueous phase, and the polarity of the respective phases. Protein adsorption at fluid interfaces plays a critical role in the production and stability of many food emulsions and foams like mayonnaise, whipped cream, or meringue and in physiological fluids like tear film, lipid droplets, or pulmonary surfactant. Certain enzymes like lipases involved in fat digestion act by adsorbing at the oil-water interface of ingested fat. Some animals exploit the foaming of secreted proteins, such as the sea snail Janthina janthina for passive flotation or certain species of spittlebug (Cercopidae) nymphs for protection from predators, moisture loss, and UV-radiation.

5.5.9 Protein Adsorption Prediction

Protein adsorption on solid surfaces plays an important role in many applications, including medical and dental practice (e.g., implants and catheters), biomedical research (e.g., drug delivery and release), and devices for diagnostics and drug discovery (e.g., assays, microarrays, and lab-on-a-chip).

Therefore, the accurate prediction of protein adsorption is critical in the progression of these sectors. Unfortunately, due to the inherent complexity of both the adsorption process and the protein molecular surface in general, protein adsorption prediction has continued to frustrate researchers. However, the Protein Adsorption Predictor is an application which aims to forecast protein concentration on surfaces using semi-empirical relationships.

Biomolecular Adsorption Database

In an effort to aid in the prediction of protein adsorption, researchers have created the Biomolecular Adsorption Database (BAD). BAD is a freely available online database with experimental protein adsorption data collected from the literature. The database can be used for the selection of materials for microfluidic device fabrication and for the selection of optimum operating conditions of lab-on-a-chip devices. The first version was published in 2009. In 2024, an upgraded version, BAD2.0, was made publically available here. This database comprises 865 protein adsorption records (flat solid surfaces) from peer-reviewed literature. BAD 2.0 comprises only data that quantitatively report, completely, the descriptors of all three classes of descriptors relevant to protein adsorption:

- The adsorbed protein, preferably identifiable in the Protein Data Bank (PDB); to allow further derivation of protein descriptors;
- the flat surface (type, water contact angle or surface tension); and
- fluid descriptors, i.e., protein concentration in solution, pH, ionic strength, and temperature.

BAD2.0 also comprises auxiliary data, i.e., method of measurement, and the DOI of the relevant references for easy retrieval.

5.6 ADSORPTION CHILLERS

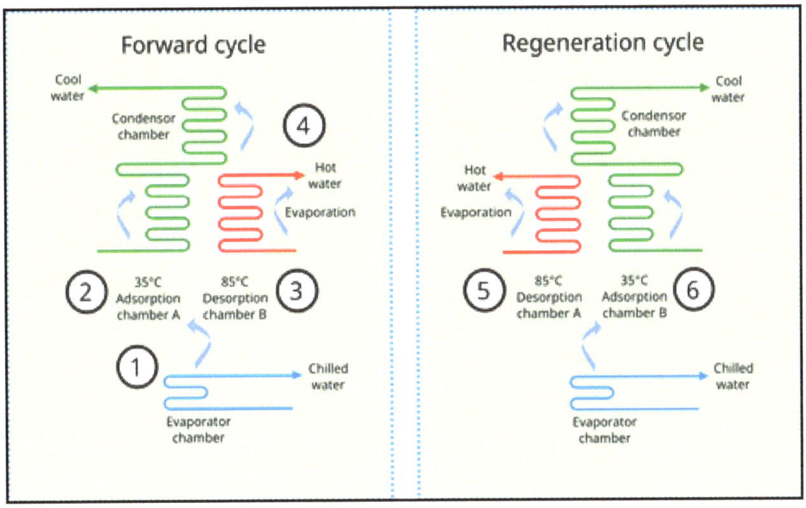

Fig: A Schematic Diagram of an Adsorption Chiller

(1) Heat is lost through evaporation of refrigerant, (2) refrigerant vapour is adsorbed onto the solid medium, (3) refrigerant is desorbed from the solid medium section not in use, (4) refrigerant is condensed and returned to the start, (5) & (6) solid medium is cycled between adsorption and desorption to regenerate it.

Combining an adsorbent with a refrigerant, adsorption chillers use heat to provide a cooling effect. This heat, in the form of hot water, may come from any number of industrial sources including waste heat from industrial processes, prime heat from solar thermal installations or from the exhaust or water jacket heat of a piston engine or turbine.

Although there are similarities between adsorption chillers and absorption refrigeration, the former is based on the interaction between gases and solids. The adsorption chamber of the chiller is filled with a solid material (for example zeolite, silica gel, alumina, active carbon or certain types of metal salts), which in its neutral state has adsorbed the refrigerant. When heated, the solid desorbs (releases) refrigerant vapour, which subsequently is cooled and liquefied. This liquid refrigerant then provides a cooling effect at the evaporator from its enthalpy of vaporization. In the final stage the refrigerant vapour is (re)adsorbed into the solid. As an adsorption chiller requires no compressor, it is relatively quiet.

5.7 PORTAL SITE MEDIATED ADSORPTION

Portal site mediated adsorption is a model for site-selective activated gas adsorption in metallic catalytic systems that contain a variety of different adsorption sites. In such systems, low-coordination "edge and corner" defect-like sites can exhibit significantly lower adsorption enthalpies than high-coordination (basal plane) sites. As a result, these sites can serve as "portals" for very rapid adsorption to the rest of the surface. The phenomenon relies on the common "spillover" effect, where certain adsorbed species exhibit high mobility on some surfaces. The model explains seemingly inconsistent observations of gas adsorption thermodynamics and kinetics in catalytic systems where surfaces can exist in a range of coordination structures, and it has been successfully applied to bimetallic catalytic systems where synergistic activity is observed.

In contrast to pure spillover, portal site adsorption refers to surface diffusion to adjacent adsorption sites, not to non-adsorptive support surfaces. The model appears to have been first proposed for carbon monoxide on silica-supported platinum by Brandt et al. (1993).

A similar, but independent model was developed by King and co-workers to describe hydrogen adsorption on silica-supported alkali promoted ruthenium, silver-ruthenium and copper-ruthenium bimetallic catalysts. The same group applied the model to CO hydrogenation (Fischer–Tropsch synthesis). Zupanc et al. (2002) subsequently confirmed the same model for hydrogen adsorption on magnesia-supported caesium-ruthenium bimetallic catalysts. Trens et al. (2009) have similarly described CO surface diffusion on carbon-supported Pt particles of varying morphology.

5.8 ADSORPTION SPILLOVER

In the case catalytic or adsorbent systems where a metal species is dispersed upon a support (or carrier) material (often quasi-inert oxides, such as alumina or silica), it is possible for an adsorptive species to indirectly adsorb to the support surface under conditions where such adsorption is thermodynamically unfavorable. The presence of the metal serves as a lower-energy pathway for gaseous species to first adsorb to the metal and then diffuse on the support surface. This is possible because the adsorbed species attains a lower energy state once it has adsorbed to the metal, thus lowering the activation barrier between the gas phase species and the support-adsorbed species.

Hydrogen spillover is the most common example of an adsorptive spillover. In the case of hydrogen, adsorption is most often accompanied with dissociation of molecular hydrogen (H_2) to atomic hydrogen (H), followed by spillover of the hydrogen atoms present. The spillover effect has been used to explain many observations in heterogeneous catalysis and adsorption.

5.9 POLYMER ADSORPTION

Adsorption of molecules onto polymer surfaces is central to a number of applications, including development of non-stick coatings and in various biomedical devices. Polymers may also be adsorbed to surfaces through polyelectrolyte adsorption.

5.10 IN VIRUSES

Adsorption is the first step in the viral life cycle. The next steps are penetration, uncoating, synthesis (transcription if needed, and translation), and release. The virus replication cycle, in this respect, is similar for all types of viruses.

Factors such as transcription may or may not be needed if the virus is able to integrate its genomic information in the cell's nucleus, or if the virus can replicate itself directly within the cell's cytoplasm.

CHAPTER

6 Extraction

Extraction in chemistry is a separation process consisting of the separation of a substance from a matrix. The distribution of a solute between two phases is an equilibrium condition described by partition theory. This is based on exactly how the analyte moves from the initial solvent into the extracting solvent. The term washing may also be used to refer to an extraction in which impurities are extracted from the solvent containing the desired compound.

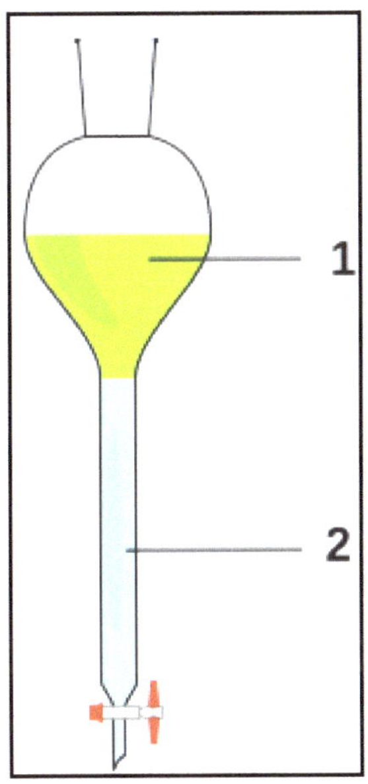

Fig: Schematic of a Separatory Funnel Showing Two Immiscible Liquids, where 1 is Any Phase Less Dense than 2. Phase 1 is Typically an Organic Solvent and 2 an Aqueous Phase

6.1 LABORATORY APPLICATIONS AND EXAMPLES

Liquid-liquid extractions in the laboratory usually make use of a separatory funnel, where two immiscible phases are combined to separate a solute from one phase into the other, according to the relative solubility in each of the phases. Typically, this will be to extract organic compounds out of an aqueous phase and into an organic phase, but may also include extracting water-soluble impurities from an organic phase into an aqueous phase.

Common extractants may be arranged in increasing order of polarity according to the Hildebrand solubility parameter: ethyl acetate < acetone < ethanol < methanol < acetone:water (7:3) < ethanol:water (8:2) < methanol:water (8:2) < water . Solid-liquid extractions at laboratory scales can use Soxhlet extractors. A solid sample containing the desired compound along with impurities is placed in the thimble. An extracting solvent is chosen in which the impurities are insoluble and the desired compound has at least limited solubility.

The solvent is refluxed and condensed solvent falls into the thimble and dissolves the desired compound which then passes back through the filter into the flask. After extraction is complete the solvent can be removed and the desired product collected.

Fig: Laboratory-Scale Liquid-Liquid Extraction

6.2 LIQUID–LIQUID EXTRACTION

Liquid–liquid extraction, also known as solvent extraction and partitioning, is a method to separate compounds or metal complexes, based on their relative solubilities in two different immiscible liquids, usually water (polar) and an organic solvent (non-polar). There is a net transfer of one or more species from one liquid into another liquid phase, generally from aqueous to organic. The transfer is driven by chemical potential, i.e. once the transfer is complete, the overall system of chemical components that make up the solutes and the solvents are in a more stable configuration (lower free energy). The solvent that is enriched in solute(s) is called extract. The feed solution that is depleted in solute(s) is called the raffinate.

Liquid–liquid extraction is a basic technique in chemical laboratories, where it is performed using a variety of apparatus, from separatory funnels to countercurrent distribution equipment called as mixer settlers. This type of process is commonly performed after a chemical reaction as part of the work-up, often including an acidic work-up.

The term partitioning is commonly used to refer to the underlying chemical and physical processes involved in liquid–liquid extraction, but on another reading may be fully synonymous with it. The term solvent extraction can also refer to the separation of a substance from a mixture by preferentially dissolving that substance in a suitable solvent. In that case, a soluble compound is separated from an insoluble compound or a complex matrix.

From a hydrometallurgical perspective, solvent extraction is exclusively used in separation and purification of uranium and plutonium, zirconium and hafnium, separation of cobalt and nickel, separation and purification of rare earth elements etc., its greatest advantage being its ability to selectively separate out even very similar metals. One obtains high-purity single metal streams on 'stripping' out the metal value from the 'loaded' organic wherein one can precipitate or deposit the metal value. Stripping is the opposite of extraction: Transfer of mass from organic to aqueous phase.

Liquid–liquid extraction is also widely used in the production of fine organic compounds, the processing of perfumes, the production of vegetable oils and biodiesel, and other industries. It is among the most common initial separation techniques, though some difficulties result in extracting out closely related functional groups.

Liquid-Liquid extraction can be substantially accelerated in microfluidic devices, reducing extraction and separation times from minutes/hours to mere seconds compared to conventional extractors.

Liquid–liquid extraction is possible in non-aqueous systems: In a system consisting of a molten metal in contact with molten salts, metals can be extracted from one phase to the other. This is related to a mercury electrode where a metal can be reduced, the metal will often then dissolve in the mercury to form an amalgam that modifies its electrochemistry greatly. For example, it is possible for sodium cations to be reduced at a mercury cathode to form sodium amalgam, while at an inert electrode (such as platinum) the sodium cations are not reduced. Instead, water is reduced to hydrogen. A detergent or fine solid can be used to stabilize an emulsion, or third phase.

6.2.1 Measures of Effectiveness

Distribution Ratio

In solvent extraction, a distribution ratio (D) is often quoted as a measure of how well-extracted a species is. The distribution ratio is a measure of the total concentration of a solute in the organic phase divided by its concentration in the aqueous phase. The partition or distribution coefficient (K_d) is the ration of solute concentration in each layer upon reaching equilibrium. This distinction between D and K_d is important. The partition coefficient is a thermodynamic equilibrium constant and has a fixed value for the solute's partitioning between the two phases. The distribution ratio's value, however, changes with solution conditions if the relative amounts of A and B change. If we know the solute's equilibrium reactions within each phase and between the two phases, we can derive an algebraic relationship between K_d and D. The partition coefficient and the distribution ratio are identical if the solute has only one chemical form in each phase; however, if the solute exists in more than one chemical form in either phase, then K_d and D usually have different values. Depending on the system, the distribution ratio can be a function of temperature, the concentration of chemical species in the system, and a large number of other parameters. Note that D is related to the Gibbs Free Energy (ΔG) of the extraction process.

In solvent extraction, two immiscible liquids are shaken together. The more polar solutes dissolve preferentially in the more polar solvent, and the less polar solutes in the less polar solvent. In this experiment, the nonpolar halogens preferentially dissolve in the non-polar mineral oil.

Separation Factors

The separation factor is one distribution ratio divided by another; it is a measure of the ability of the system to separate two solutes. For instance, if the distribution ratio for nickel (D_{Ni}) is 10 and the distribution ratio for silver (D_{Ag}) is 100, then the silver/nickel separation factor ($SF_{Ag/Ni}$) is equal to $D_{Ag}/D_{Ni} = SF_{Ag/Ni} = 10$.

Measures of Success

Success of liquid–liquid extraction is measured through separation factors and decontamination factors. The best way to understand the success of an extraction column is through the liquid–liquid equilibrium (LLE) data set. The data set can then be converted into a curve to determine the steady state partitioning behavior of the solute between the two phases. The y-axis is the concentration of solute in the extract (solvent) phase, and the x-axis is the concentration of the solute in the raffinate phase. From here, one can determine steps for optimization of the process.

6.2.2 Techniques

Batch Wise Single Stage Extractions

This is commonly used on the small scale in chemical labs. It is normal to use a separating funnel. Processes include DLLME and direct organic extraction. After equilibration, the extract phase containing the desired solute is separated out for further processing.

Dispersive Liquid–Liquid Microextraction (DLLME)

A process used to extract small amounts of organic compounds from water samples. This process is done by injecting small amounts of an appropriate extraction solvent (C_2Cl_4) and a disperser solvent (acetone) into the aqueous solution. The resulting solution is then centrifuged to separate the organic and aqueous layers. This process is useful in extraction organic compounds such as organochloride and organophsophorus pesticides, as well as substituted benzene compounds from water samples.

Direct Organic Extraction

By mixing partially organic soluble samples in organic solvent (toluene, benzene, xylene), the organic soluble compounds will dissolve into the solvent and can be separated using a separatory funnel. This process is valuable in the extraction of proteins and specifically phosphoprotein and phosphopeptide phosphatases.

Another example of this application is extracting anisole from a mixture of water and 5% acetic acid using ether, then the anisole will enter the organic phase. The two phases would then be separated. The acetic acid can then be scrubbed (removed) from the organic phase by shaking the organic extract with sodium bicarbonate. The acetic acid reacts with the sodium bicarbonate to form sodium acetate, carbon dioxide, and water.

Caffeine can also be extracted from coffee beans and tea leaves using a direct organic extraction. The beans or leaves can be soaked in ethyl acetate which favorably dissolves the caffeine, leaving a majority of the coffee or tea flavor remaining in the initial sample.

Multistage countercurrent Continuous Processes

These are commonly used in industry for the processing of metals such as the lanthanides; because the separation factors between the lanthanides are so small many extraction stages are needed. In the multistage processes, the aqueous raffinate from one extraction unit is fed to the next unit as the aqueous feed, while the organic phase is moved in the opposite direction. Hence, in this way, even if the separation between two metals in each stage is small, the overall system can have a higher decontamination factor.

Multistage countercurrent arrays have been used for the separation of lanthanides. For the design of a good process, the distribution ratio should be not too high (>100) or too low (<0.1) in the extraction portion of the process. It is often the case that the process will have a section for scrubbing unwanted metals from the organic phase, and finally a stripping section to obtain the metal back from the organic phase.

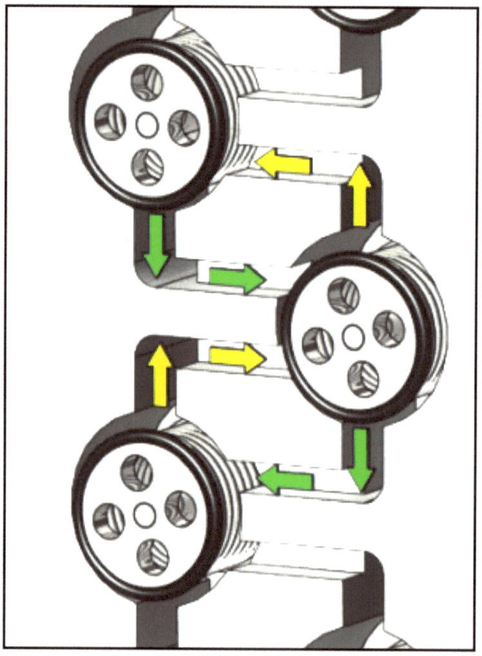

Fig: Coflore Continuous Countercurrent Extractor

Mixer–Settlers

Battery of mixer-settlers counter currently interconnected. Each mixer-settler unit provides a single stage of extraction. A mixer settler consists of a first stage that mixes the phases together followed by a quiescent settling stage that allows the phases to separate by gravity.

A novel settling device, Sudhin BioSettler, can separate an oil-water emulsion continuously at a much faster rate than simple gravity settlers. In this photo, an oil-water emulsion, stirred by an impeller in an external reservoir and pumped continuously into the two bottom side ports of BioSettler, is separated very quickly into a clear organic (mineral oil) layer exiting via the top of BioSettler and an aqueous (coloured with a red food dye) layer being pumped out continuously from the bottom of BioSettler.

In the multistage countercurrent process, multiple mixer settlers are installed with mixing and settling chambers located at alternating ends for each stage (since the outlet of the settling sections feed the inlets of the adjacent stage's mixing sections). Mixer-settlers are used when a process requires longer residence times and when the solutions are easily separated by gravity. They require a large facility footprint, but do not require much headspace, and need limited remote maintenance capability for occasional replacement of mixing motors. (Colven, 1956; Davidson, 1957)

Fig: Continuous Separation of Oil and Water Mixture

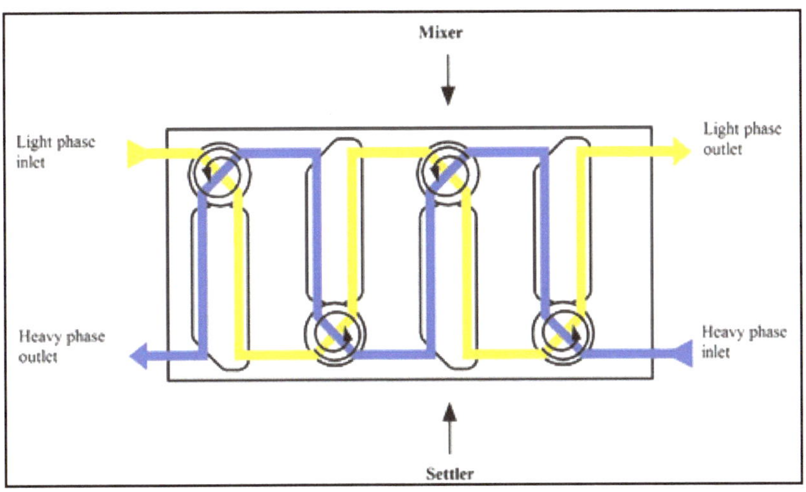

Fig: Four Stage Battery of Mixer-Settlers for Counter-Current Extraction

Centrifugal Extractors

Centrifugal extractors mix and separate in one unit. Two liquids will be intensively mixed between the spinning rotor and the stationary housing at speeds up to 6000 RPM. This develops great surfaces for an ideal mass transfer from the aqueous phase into the organic phase. At 200–2000 g, both phases will be separated again. Centrifugal extractors minimize the solvent in the process, optimize the product load in the solvent and extract the aqueous phase completely. Counter current and cross current extractions are easily established.

Extraction without Chemical Change

Some solutes such as noble gases can be extracted from one phase to another without the need for a chemical reaction. This is the simplest type of solvent extraction. When a solvent is extracted, two immiscible liquids are shaken together. The more polar solutes dissolve preferentially in the more polar solvent, and the less polar solutes in the less polar solvent.

Some solutes that do not at first sight appear to undergo a reaction during the extraction process do not have distribution ratio that is independent of concentration. A classic example is the extraction of carboxylic acids (HA) into nonpolar media such as benzene. Here, it is often the case that the carboxylic acid will form a dimer in the organic layer so the distribution ratio will change as a function of the acid concentration (measured in either phase).

For this case, the extraction constant k is described by $k = [HA_{organic}]^2/[HA_{aqueous}]$

Solvation Mechanism

Using solvent extraction it is possible to extract uranium, plutonium, thorium and many rare earth elements from acid solutions in a selective way by using the right choice of organic extracting solvent and diluent. One solvent used for this purpose is the organophosphate tributyl phosphate (TBP). The PUREX process that is commonly used in nuclear reprocessing uses a mixture of tri-n-butyl phosphate and an inert hydrocarbon (kerosene), the uranium(VI) are extracted from strong nitric acid and are back-extracted (stripped) using weak nitric acid. An organic soluble uranium complex $[UO_2(TBP)_2(NO_3)_2]$ is formed, then the organic layer bearing the uranium is brought into contact with a dilute nitric acid solution; the equilibrium is shifted away from the organic soluble uranium complex and towards the free TBP and uranyl nitrate in dilute nitric acid.

The plutonium(IV) forms a similar complex to the uranium(VI), but it is possible to strip the plutonium in more than one way; a reducing agent that converts the plutonium to the trivalent oxidation state can be added. This oxidation state does not form a stable complex with TBP and nitrate unless the nitrate concentration is very high (circa 10 mol/L nitrate is required in the aqueous phase). Another method is to simply use dilute nitric acid as a stripping agent for the plutonium. This PUREX chemistry is a classic example of a solvation extraction. In this case, $D_U = k\,[TBP]^2[NO_3^-]^2$.

Ion Exchange Mechanism

Another extraction mechanism is known as the ion exchange mechanism. Here, when an ion is transferred from the aqueous phase to the organic phase, another ion is transferred in the other direction to maintain the charge balance.

This additional ion is often a hydrogen ion; for ion exchange mechanisms, the distribution ratio is often a function of pH. An example of an ion exchange extraction would be the extraction of americium by a combination of terpyridine and a carboxylic acid in tert-butyl benzene. In this case

$$D_{Am} = k\,[\text{terpyridine}]^1[\text{carboxylic acid}]^3[H^+]^{-3}$$

Another example is the extraction of zinc, cadmium, or lead by a dialkyl phosphinic acid (R_2PO_2H) into a nonpolar diluent such as an alkane. A non-polar diluent favours the formation of uncharged non-polar metal complexes.

Some extraction systems are able to extract metals by both the solvation and ion exchange mechanisms; an example of such a system is the americium (and lanthanide) extraction from nitric acid by a combination of 6,6'-bis-(5,6-dipentyl-1,2,4-triazin-3-yl)-2,2'-bipyridine and 2-bromohexanoic acid in tert-butyl benzene. At both high- and low-nitric acid concentrations, the metal distribution ratio is higher than it is for an intermediate nitric acid concentration.

Ion Pair Extraction

It is possible by careful choice of counterion to extract a metal. For instance, if the nitrate concentration is high, it is possible to extract americium as an anionic nitrate complex if the mixture contains a lipophilic quaternary ammonium salt.

An example that is more likely to be encountered by the 'average' chemist is the use of a phase transfer catalyst. This is a charged species that transfers another ion to the organic phase. The ion reacts and then forms another ion, which is then transferred back to the aqueous phase.

For instance, the 31.1 kJ mol^{-1} is required to transfer an acetate anion into nitrobenzene, while the energy required to transfer a chloride anion from an aqueous phase to nitrobenzene is 43.8 kJ mol^{-1}.

Hence, if the aqueous phase in a reaction is a solution of sodium acetate while the organic phase is a nitrobenzene solution of benzyl chloride, then, when a phase transfer catalyst, the acetate anions can be transferred from the aqueous layer where they react with the benzyl chloride to form benzyl acetate and a chloride anion. The chloride anion is then transferred to the aqueous phase. The transfer energies of the anions contribute to that given out by the reaction.

A 43.8 to 31.1 kJ mol^{-1} = 12.7 kJ mol^{-1} of additional energy is given out by the reaction when compared with energy if the reaction had been done in nitrobenzene using one equivalent weight of a tetraalkylammonium acetate.

Types of Aqueous Two-Phase Extractions

Polymer–polymer systems. In a Polymer–polymer system, both phases are generated by a dissolved polymer. The heavy phase will generally be a polysaccharide, and the light phase is generally Polyethylene glycol (PEG). Traditionally, the polysaccharide used is dextran. However, dextran is relatively expensive, and research has been exploring using less expensive polysaccharides to generate the heavy phase. If the target compound being separated is a protein or enzyme, it is possible to incorporate a ligand to the target into one of the polymer phases. This improves the target's affinity to that phase, and improves its ability to partition from one phase into the other. This, as well as the absence of solvents or other denaturing agents, makes polymer–polymer extractions an attractive option for purifying proteins. The two phases of a polymer–polymer system often have very similar densities, and very low surface tension between them. Because of this, demixing a polymer–polymer system is often much more difficult than demixing a solvent extraction. Methods to improve the demixing include centrifugation, and application of an electric field.

Polymer–salt systems. Aqueous two-phase systems can also be generated by generating the heavy phase with a concentrated salt solution. The polymer phase used is generally still PEG. Generally, a kosmotropic salt, such as Na_3PO_4 is used, however PEG–NaCl systems have been documented when the salt concentration is high enough. Since polymer–salt systems demix readily they are easier to use. However, at high salt concentrations, proteins generally either denature, or precipitate from solution. Thus, polymer–salt systems are not as useful for purifying proteins.

Ionic liquids systems. Ionic liquids are ionic compounds with low melting points. While they are not technically aqueous, recent research has experimented with using them in an extraction that does not use organic solvents.

DNA Purification

The ability to purify DNA from a sample is important for many modern biotechnology processes. However, samples often contain nucleases that degrade the target DNA before it can be purified. It has been shown that DNA fragments will partition into the light phase of a polymer–salt separation system. If ligands known to bind and deactivate nucleases are incorporated into the polymer phase, the nucleases will then partition into the heavy phase and be deactivated. Thus, this polymer–salt system is a useful tool for purifying DNA from a sample while simultaneously protecting it from nucleases.

Food Industry

The PEG–NaCl system has been shown to be effective at partitioning small molecules, such as peptides and nucleic acids. These compounds are often flavorants or odorants. The system could then be used by the food industry to isolate or eliminate particular flavors. Caffeine extraction used to be done using liquid–liquid extraction, specifically direct and indirect liquid–liquid extraction (Swiss Water Method), but has since moved towards super-critical CO_2 as it is cheaper and can be done on a commercial scale.

Analytical Chemistry

Often there are chemical species present or necessary at one stage of sample processing that will interfere with the analysis. For example, some air monitoring is performed by drawing air through a small glass tube filled with sorbent particles that have been coated with a chemical to stabilize or derivatize the analyte of interest. The coating may be of such a concentration or characteristics that it would damage the instrumentation or interfere with the analysis. If the sample can be extracted from the sorbent using a nonpolar solvent (such as toluene or carbon disulfide), and the coating is polar (such as HBr or phosphoric acid) the dissolved coating will partition into the aqueous phase. Clearly the reverse is true as well, using polar extraction solvent and a nonpolar solvent to partition a nonpolar interferent. A small aliquot of the organic phase (or in the latter case, polar phase) can then be injected into the instrument for analysis.

Purification of Amines

Amines (analogously to ammonia) have a lone pair of electrons on the nitrogen atom that can form a relatively weak bond to a hydrogen atom. It is therefore the case that under acidic conditions amines are typically protonated, carrying a positive charge and under basic conditions they are typically deprotonated and neutral. Amines of sufficiently low molecular weight are rather polar and can form hydrogen bonds with water and therefore will readily dissolve in aqueous solutions. Deprotonated amines on the other hand, are neutral and have greasy, nonpolar organic substituents, and therefore have a higher affinity for nonpolar inorganic solvents. As such purification steps can be carried out where an aqueous solution of an amine is neutralized with a base such as sodium hydroxide, then shaken in a separatory funnel with a nonpolar solvent that is immiscible with water. The organic phase is then drained off. Subsequent processing can recover the amine by techniques such as recrystallization, evaporation or distillation; subsequent extraction back to a polar phase can be performed by adding HCl and shaking again in a separatory funnel (at which point the ammonium ion could be recovered by adding an insoluble counterion), or in either phase, reactions could be performed as part of a chemical synthesis.

Temperature Swing Solvent Extraction

Temperature swing solvent extraction is an experimental technique for the desalination of drinking water. It has been used to remove up to 98.5% of the salt content in water, and is able to process hypersaline brines that cannot be desalinated using reverse osmosis.

6.2.3 Kinetics of Extraction

It is important to investigate the rate at which the solute is transferred between the two phases, in some cases by an alteration of the contact time it is possible to alter the selectivity of the extraction. For instance, the extraction of palladium or nickel can be very slow because the rate of ligand exchange at these metal centers is much lower than the rates for iron or silver complexes.

6.2.4 Aqueous Complexing Agents

If a complexing agent is present in the aqueous phase then it can lower the distribution ratio. For instance, in the case of iodine being distributed between water and an inert organic solvent such as carbon tetrachloride then the presence of iodide in the aqueous phase can alter the extraction chemistry: instead of $D_{I^{+2}}$ being a constant it becomes

$$D_{I^{+2}} = k[I_2\,(\text{organic})]/[I_2\,(\text{aq})][I^-\,(\text{aq})]$$

This is because the iodine reacts with the iodide to form I_3^-. The I_3^- anion is an example of a polyhalide anion that is quite common.

6.2.5 Industrial Process Design

In a typical scenario, an industrial process will use an extraction step in which solutes are transferred from the aqueous phase to the organic phase; this is often followed by a scrubbing stage in which unwanted solutes are removed from the organic phase, then a stripping stage in which the wanted solutes are removed from the organic phase. The organic phase may then be treated to make it ready for use again. After use, the organic phase may be subjected to a cleaning step to remove any degradation products; for instance, in PUREX plants, the used organic phase is washed with sodium carbonate solution to remove any dibutyl hydrogen phosphate or butyl dihydrogen phosphate that might be present.

Liquid-Liquid Equilibrium Calculations

In order to calculate the phase equilibrium, it is necessary to use a thermodynamic model such as NRTL, UNIQUAC, etc. The corresponding parameters of these models can be obtained from literature (e.g. Dechema Chemistry Data Series, Dortmund Data Bank, etc.) or by a correlation process of experimental data.

6.2.6 Equipment

While solvent extraction is often done on a small scale by synthetic lab chemists using a separatory funnel, Craig apparatus or membrane-based techniques, it is normally done on the industrial scale using machines that bring the two liquid phases into contact with each other. Such machines include centrifugal contactors, Thin Layer Extraction, spray columns, pulsed columns, and mixer-settlers.

6.2.7 Extraction of Metals

- **Cobalt:** The extraction of cobalt from hydrochloric acid using Alamine 336 (tri-octyl/decyl amine) in meta-xylene. Cobalt can be extracted also using Ionquest 290 or Cyanex 272 {bis-(2,4,4-trimethylpentyl) phosphinic acid}.
- **Copper:** Copper can be extracted using hydroxyoximes as extractants, a recent paper describes an extractant that has a good selectivity for copper over cobalt and nickel.
- **Neodymium:** The rare earth element Neodymium is extracted by di(2-ethyl-hexyl)phosphoric acid into hexane by an ion exchange mechanism.
- **Nickel:** Nickel can be extracted using di(2-ethyl-hexyl)phosphoric acid and tributyl phosphate in a hydrocarbon diluent (Shellsol).
- **Palladium and Platinum:** Dialkyl sulfides, tributyl phosphate and alkyl amines have been used for extracting palladium and platinum.
- **Polonium:** Polonium is produced in reactors from natural ^{209}Bi, bombarded with neutrons, creating ^{210}Bi, which then decays to ^{210}Po via beta-minus decay. The final purification is done pyrochemically followed by liquid-liquid extraction vs sodium hydroxide at 500 °C.
- **Zinc and Cadmium:** Zinc and cadmium are both extracted by an ion exchange process, the N,N,N′,N′-tetrakis(2-pyridylmethyl) ethylenediamine (TPEN) acts as a masking agent for the zinc and an extractant for the cadmium. In the modified Zincex process, zinc is separated from most divalent ions by solvent extraction. D2EHPA (Di (2) ethyl hexyl phosphoric acid) is used for this. A zinc ion replaces the proton from two D2EHPA molecules. To strip the zinc from the D2EHPA,

sulfuric acid is used, at a concentration of above 170g/L (typically 240-265g/L).

- **Lithium:** Lithium extraction is more popular due to the high demand of lithium-ion batteries. TBP (Tri-butyl phosphate) and $FeCl_3$ are mostly used to extract lithium from brine (with high Li/Mg ratio). Alternatively, Cyanex 272 was also used to extract lithium. The mechanism of lithium extraction was found differently from other metals, such as cobalt, due to the weak coordinating bonding between lithium ions and extractants.

CHAPTER 7

Membrane Separation Processes

Membrane technology encompasses the scientific processes used in the construction and application of membranes. Membranes are used to facilitate the transport or rejection of substances between mediums, and the mechanical separation of gas and liquid streams. In the simplest case, filtration is achieved when the pores of the membrane are smaller than the diameter of the undesired substance, such as a harmful microorganism. Membrane technology is commonly used in industries such as water treatment, chemical and metal processing, pharmaceuticals, biotechnology, the food industry, as well as the removal of environmental pollutants.

After membrane construction, there is a need to characterize the prepared membrane to know more about its parameters, like pore size, function group, material properties, etc., which are difficult to determine in advance. In this process, instruments such as the Scanning Electron Microscope, the Transmission electron Microscope, the Fourier Transform Infrared Spectroscopy, X-ray Diffraction, and Liquid–Liquid Displacement Porosimetry are utilized.

7.1 INTRODUCTION

Membrane technology covers all engineering approaches for the transport of substances between two fractions with the help of semi-permeable membranes. In general, mechanical separation processes for separating gaseous or liquid streams use membrane technology. In recent years, different methods have been used to remove environmental pollutants, like adsorption, oxidation, and membrane separation. Different pollution occurs in the environment like air pollution, waste water pollution etc. As per industry requirement to prevent industrial pollution because more than 70% of environmental pollution occurs due to industries. It is their responsibility to follow government rules of the Air Pollution Control & Prevention Act 1981 to maintain and prevent the harmful chemical release into the environment. Make sure to do prevention & safety processes after that industries are able to release their waste in the environment.

Biomass-based Membrane technology is one of the most promising technologies for use as a pollutants removal weapon because it has low cost, more efficiency, &

lack of secondary pollutants. Typically polysulfone, polyvinylidene fluoride, and polypropylene are used in the membrane preparation process. These membrane materials are non-renewable and non-biodegradable which create harmful environmental pollution. Researchers are trying to find a solution to synthesize an eco-friendly membrane which avoids environmental pollution. Synthesis of biodegradable material with the help of naturally available material such as biomass-based membrane synthesis can be used to remove pollutants.

7.1.1 Membrane Overview

Fig: Ultrafiltration for a Swimming Pool

Membrane separation processes operate without heating and therefore use less energy than conventional thermal separation processes such as distillation, sublimation or crystallization. The separation process is purely physical and both fractions (permeate and retentate) can be obtained as useful products. Cold separation using membrane technology is widely used in the food technology, biotechnology and pharmaceutical industries. Furthermore, using membranes enables separations to take place that would be impossible using thermal separation methods. For example, it is impossible to separate the constituents of azeotropic liquids or solutes which form isomorphic crystals by distillation or recrystallization but such separations can be achieved using membrane technology. Depending on the type of membrane, the selective separation of certain individual substances or substance mixtures is possible. Important technical applications include the production of drinking water by reverse osmosis. In waste water treatment, membrane technology is becoming increasingly important. Ultra/microfiltration can be very effective in removing colloids and macromolecules from wastewater.

This is needed if wastewater is discharged into sensitive waters especially those designated for contact water sports and recreation.

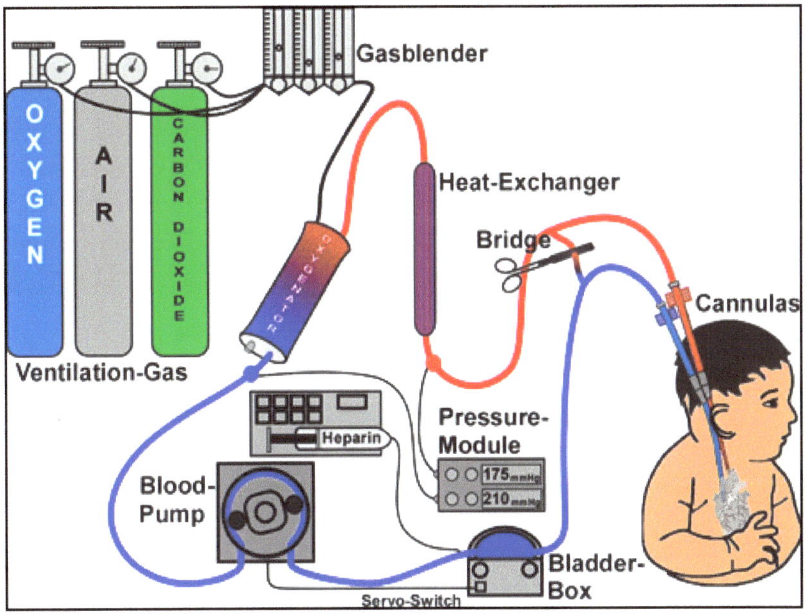

Fig: Venous-Arterial Extracorporeal Membrane Oxygenation Scheme

About half of the market is in medical applications such as artificial kidneys to remove toxic substances by hemodialysis and as artificial lung for bubble-free supply of oxygen in the blood.

The importance of membrane technology is growing in the field of environmental protection (Nano-Mem-Pro IPPC Database). Even in modern energy recovery techniques, membranes are increasingly used, for example in fuel cells and in osmotic power plants.

7.2 MASS TRANSFER

Two basic models can be distinguished for mass transfer through the membrane:

- the solution-diffusion model and
- the hydrodynamic model.

In real membranes, these two transport mechanisms certainly occur side by side, especially during ultra-filtration.

7.2.1 Solution-Diffusion Model

In the solution-diffusion model, transport occurs only by diffusion. The component that needs to be transported must first be dissolved in the membrane. The general approach of the solution-diffusion model is to assume that the chemical potential of the feed and permeate fluids are in equilibrium with the adjacent membrane surfaces such that appropriate expressions for the chemical potential in the fluid and membrane phases can be equated at the solution-membrane interface. This principle is more important for dense membranes without natural pores such as those used for reverse osmosis and in fuel cells. During the filtration process a boundary layer forms on the membrane. This concentration gradient is created by molecules which cannot pass through the membrane. The effect is referred to as concentration polarization and, occurring during the filtration, leads to a reduced trans-membrane flow (flux). Concentration polarization is, in principle, reversible by cleaning the membrane which results in the initial flux being almost totally restored. Using a tangential flow to the membrane (cross-flow filtration) can also minimize concentration polarization.

7.2.2 Hydrodynamic Model

Transport through pores – in the simplest case – is done convectively. This requires the size of the pores to be smaller than the diameter of the two separate components. Membranes that function according to this principle are used mainly in micro- and ultrafiltration. They are used to separate macromolecules from solutions, colloids from a dispersion or remove bacteria. During this process, the retained particles or molecules form a pulpy mass (filter cake) on the membrane, and this blockage of the membrane hampers the filtration. This blockage can be reduced by the use of the cross-flow method (cross-flow filtration). Here, the liquid to be filtered flows along the front of the membrane and is separated by the pressure difference between the front and back of the membrane into retentate (the flowing concentrate) on the front and permeate (filtrate) on the back. The tangential flow on the front creates a shear stress that cracks the filter cake and reduces the fouling.

7.3 MEMBRANE SHAPES AND FLOW GEOMETRIES

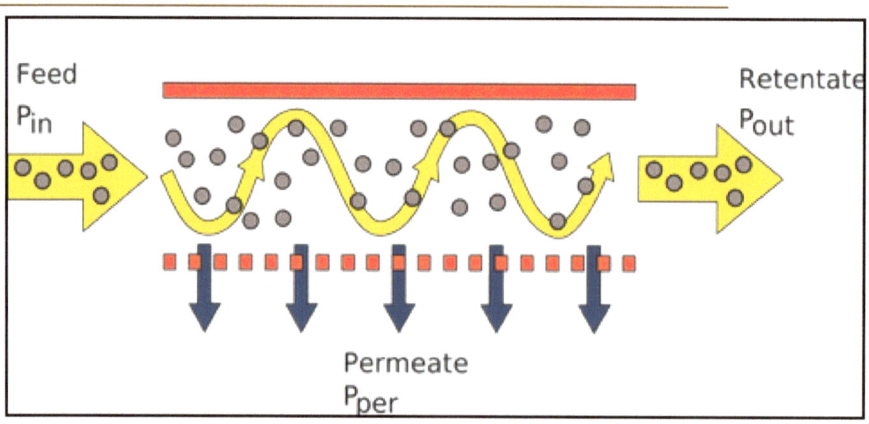

Fig: Cross-Flow Geometry

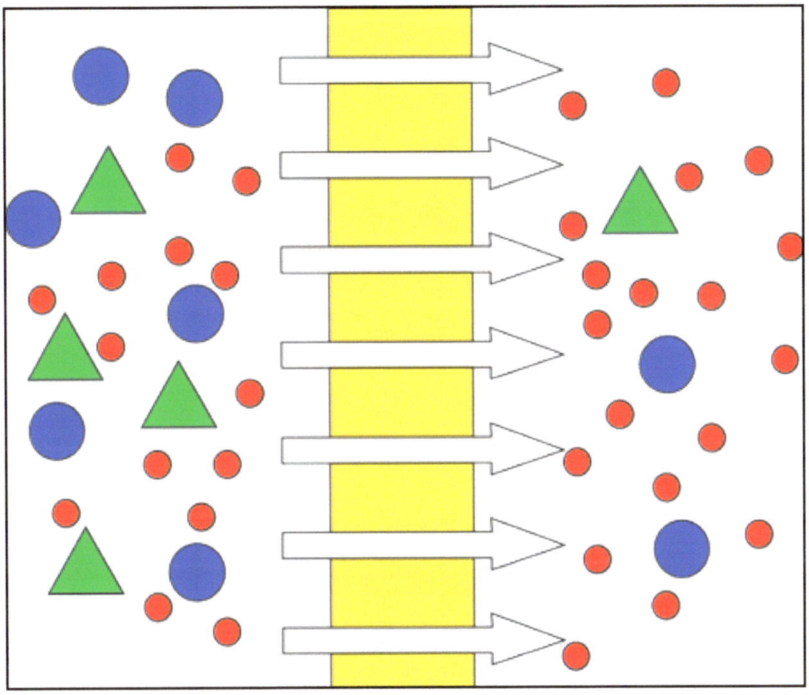

Fig: Dead-End Geometry

There are two main flow configurations of membrane processes: cross-flow (or tangential flow) and dead-end filtrations. In cross-flow filtration the feed flow is tangential to the surface of the membrane, retentate is removed from the same side further downstream, whereas the permeate flow is tracked on the other side. In dead-end filtration, the direction of the fluid flow is normal to the membrane surface. Both flow geometries offer some advantages and disadvantages. Generally, dead-end filtration is used for feasibility studies on a laboratory scale. The dead-end membranes are relatively easy to fabricate which reduces the cost of the separation process.

The dead-end membrane separation process is easy to implement and the process is usually cheaper than cross-flow membrane filtration. The dead-end filtration process is usually a batch-type process, where the filtering solution is loaded (or slowly fed) into the membrane device, which then allows passage of some particles subject to the driving force. The main disadvantage of dead-end filtration is the extensive membrane fouling and concentration polarization. The fouling is usually induced faster at higher driving forces. Membrane fouling and particle retention in a feed solution also builds up a concentration gradients and particle backflow (concentration polarization). The tangential flow devices are more cost and labor-intensive, but they are less susceptible to fouling due to the sweeping effects and high shear rates of the passing flow.

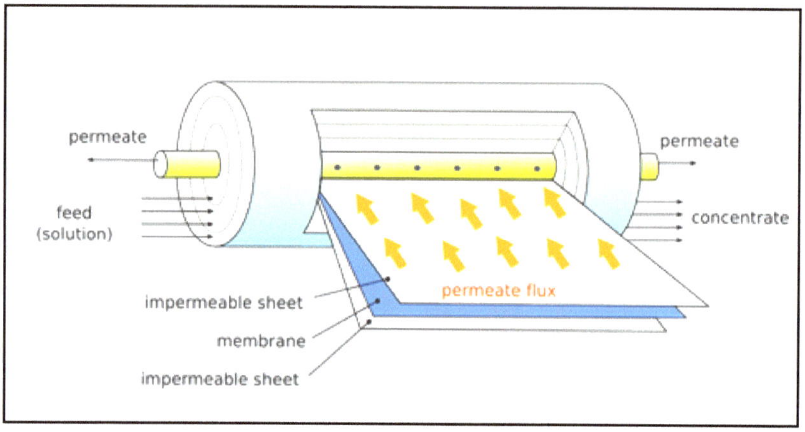

Fig: Spiral Wound Membrane Module

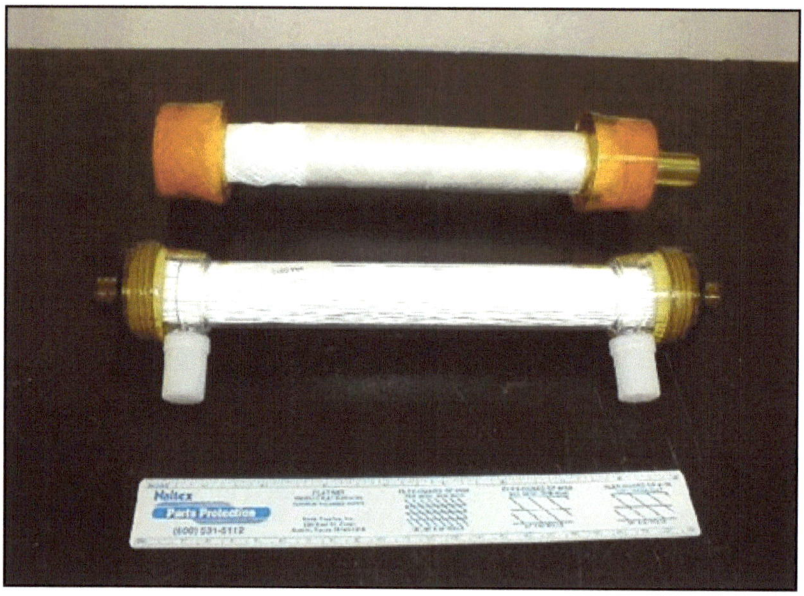

Fig: Hollow Fiber Membrane Module

The most commonly used synthetic membrane devices (modules) are flat sheets/plates, spiral wounds, and hollow fibers. Flat membranes used in filtration and separation processes can be enhanced with surface patterning, where microscopic structures are introduced to improve performance. These patterns increase surface area, optimize water flow, and reduce fouling, leading to higher permeability and longer membrane lifespan. Research has shown that such modifications can significantly enhance efficiency in water purification, energy applications, and industrial separations.

Flat plates are usually constructed as circular thin flat membrane surfaces to be used in dead-end geometry modules. Spiral wounds are constructed from similar flat membranes but in the form of a "pocket" containing two membrane sheets separated by a highly porous support plate. Several such pockets are then wound around a tube to create a tangential flow geometry and to reduce membrane fouling. Hollow fiber modules consist of an assembly of self-supporting fibers with dense skin separation layers, and a more open matrix helping to withstand pressure gradients and maintain structural integrity. The hollow fiber modules can contain up to 10,000 fibers ranging from 200 to 2500 μm in diameter; The main advantage of hollow fiber modules is the very large surface area within an enclosed volume, increasing the efficiency of the separation process.

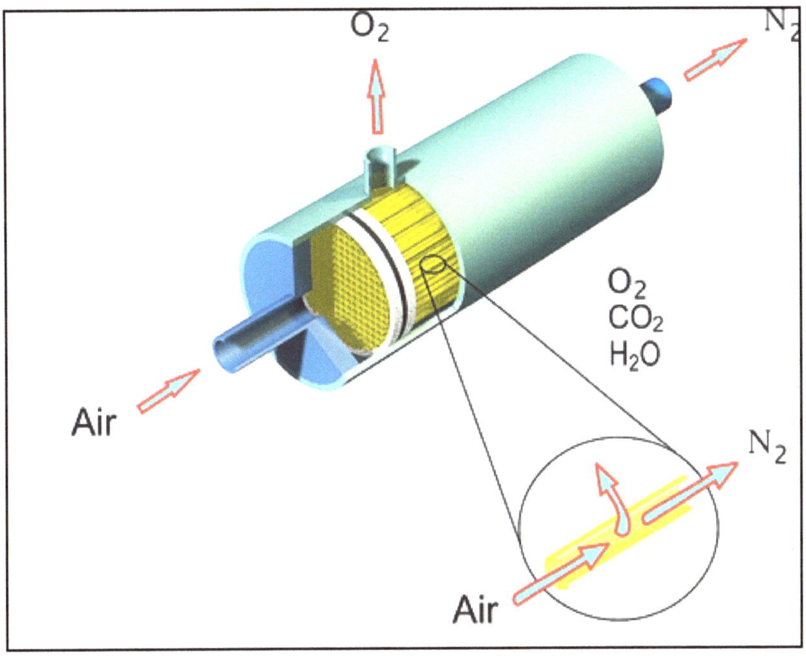

Fig: Separation of Air into Oxygen and Nitrogen through a Membrane

The Disc tube module uses a cross-flow geometry and consists of a pressure tube and hydraulic discs, which are held by a central tension rod, and membrane cushions that lie between two discs.

7.4 MEMBRANE PERFORMANCE AND GOVERNING EQUATIONS

The selection of synthetic membranes for a targeted separation process is usually based on few requirements. Membranes have to provide enough mass transfer area to process large amounts of feed stream. The selected membrane has to have high selectivity (rejection) properties for certain particles; it has to resist fouling and to have high mechanical stability. It also needs to be reproducible and to have low manufacturing costs. The main modeling equation for the dead-end filtration at constant pressure drop is represented by Darcy's law:

$$\frac{dV_p}{dt} = Q = \frac{\Delta p}{\mu} A\left(\frac{1}{R_m + R}\right)$$

where V_p and Q are the volume of the permeate and its volumetric flow rate respectively (proportional to same characteristics of the feed flow), μ is dynamic viscosity of permeating fluid, A is membrane area, R_m and R are the respective resistances of membrane and growing deposit of the foulants. R_m can be interpreted as a membrane resistance to the solvent (water) permeation. This resistance is a membrane intrinsic property and is expected to be fairly constant and independent of the driving force, Δp. R is related to the type of membrane foulant, its concentration in the filtering solution, and the nature of foulant-membrane interactions. Darcy's law allows for calculation of the membrane area for a targeted separation at given conditions. The solute sieving coefficient is defined by the equation:

$$S = \frac{C_p}{C_f}$$

where C_f and C_p are the solute concentrations in feed and permeate respectively. Hydraulic permeability is defined as the inverse of resistance and is represented by the equation:

$$L_p = \frac{J}{\Delta p}$$

where J is the permeate flux which is the volumetric flow rate per unit of membrane area. The solute sieving coefficient and hydraulic permeability allow the quick assessment of the synthetic membrane performance.

7.5 MEMBRANE SEPARATION PROCESSES

Membrane separation processes have a very important role in the separation industry. Nevertheless, they were not considered technically important until the mid-1970s. Membrane separation processes differ based on separation mechanisms and size of the separated particles. The widely used membrane processes include microfiltration, ultrafiltration, nanofiltration, reverse osmosis, electrolysis, dialysis, electrodialysis, gas separation, vapor permeation, pervaporation, membrane distillation, and membrane contactors. All processes except for pervaporation involve no phase change. All processes except electrodialysis are pressure driven. Microfiltration and ultrafiltration is widely used in food and beverage processing (beer microfiltration, apple juice ultrafiltration), biotechnological applications and pharmaceutical industry (antibiotic production, protein purification), water purification and wastewater treatment, the microelectronics industry, and others. Nanofiltration and reverse osmosis membranes are mainly used for water purification purposes. Dense membranes are utilized for gas separations (removal of CO_2 from natural gas, separating N_2 from air, organic vapor removal from air or a nitrogen stream) and sometimes in membrane distillation. The later process helps in the separation of azeotropic compositions reducing the costs of distillation processes.

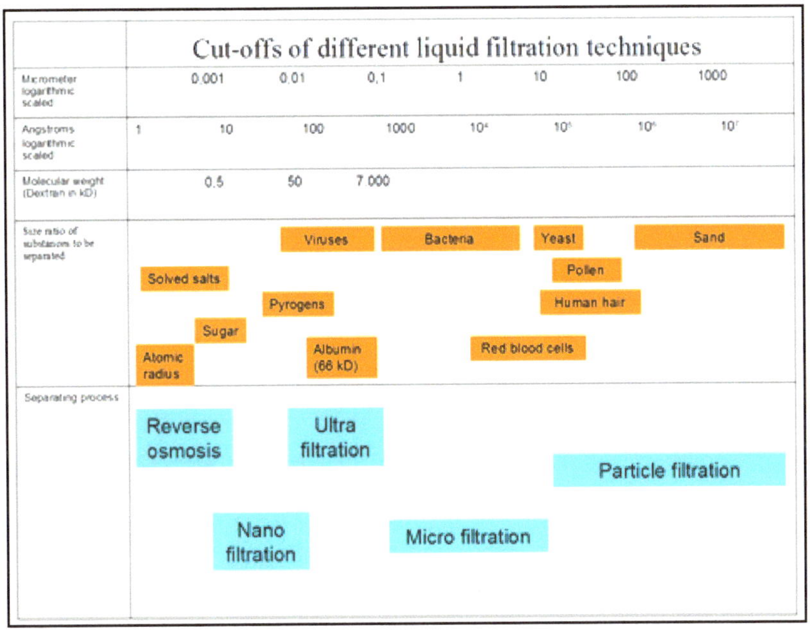

Fig: Ranges of Membrane Based Separations

7.6 PORE SIZE AND SELECTIVITY

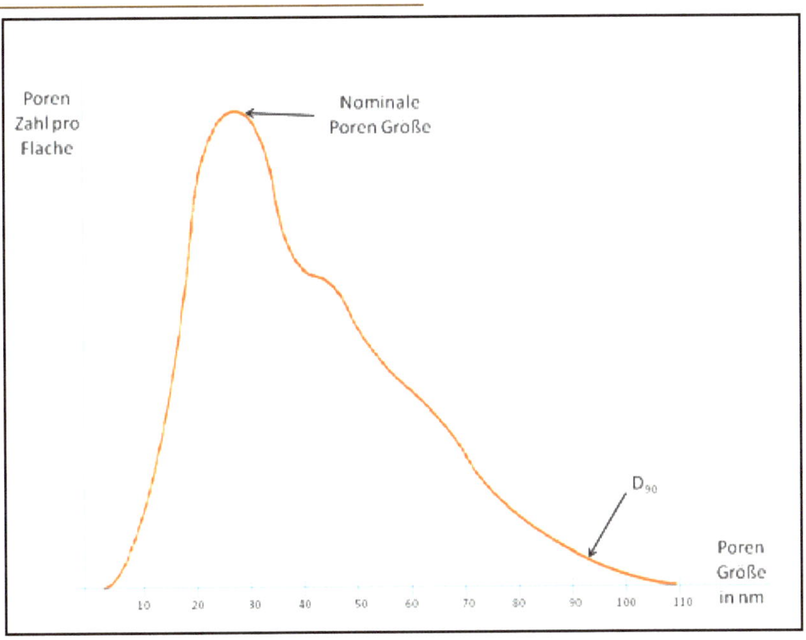

FIG: The Pore Distribution of a Fictitious Ultrafiltration Membrane with the Nominal Pore Size and the D_{90}

The pore sizes of technical membranes are specified differently depending on the manufacturer. One common distinction is by nominal pore size. It describes the maximum pore size distribution and gives only vague information about the retention capacity of a membrane. The exclusion limit or "cut-off" of the membrane is usually specified in the form of NMWC (nominal molecular weight cut-off, or MWCO, molecular weight cut off, with units in Dalton). It is defined as the minimum molecular weight of a globular molecule that is retained to 90% by the membrane. The cut-off, depending on the method, can by converted to so-called D_{90}, which is then expressed in a metric unit. In practice the MWCO of the membrane should be at least 20% lower than the molecular weight of the molecule that is to be separated.

Using track etched mica membranes Beck and Schultz demonstrated that hindered diffusion of molecules in pores can be described by the Rankin equation.

Filter membranes are divided into four classes according to pore size:

Pore Size	Molecular Mass	Process	Filtration	Removal
> 10		"Classic" filter		
> 0.1 µm	> 5000 kDa	microfiltration	< 2 bar	Of larger bacteria, yeast, particles
100-2 nm	5-5000 kDa	ultrafiltration	1-10 bar	Of bacteria, macromolecules, proteins, larger viruses
2-1 nm	0.1-5 kDa	nanofiltration	3-20 bar	Of viruses, 2-valent ions
< 1 nm	< 100 Da	reverse osmosis	10-80 bar	Of salts, small organic molecules

The form and shape of the membrane pores are highly dependent on the manufacturing process and are often difficult to specify. Therefore, for characterization, test filtrations are carried out and the pore diameter refers to the diameter of the smallest particles which could not pass through the membrane.

The rejection can be determined in various ways and provides an indirect measurement of the pore size. One possibility is the filtration of macromolecules (often dextran, polyethylene glycol or albumin), another is measurement of the cut-off by gel permeation chromatography. These methods are used mainly to measure membranes for ultrafiltration applications. Another testing method is the filtration of particles with defined size and their measurement with a particle sizer or by laser induced breakdown spectroscopy (LIBS). A vivid characterization is to measure the rejection of dextran blue or other colored molecules. The retention of bacteriophage and bacteria, the so-called "bacteria challenge test", can also provide information about the pore size.

Nominal Pore Size	Micro-Organism	ATCC Root Number
0.1 µm	*Acholeplasma laidlawii*	23206
0.3 µm	*Bacillus subtilis spores*	82
0.5 µm	*Pseudomonas diminuta*	19146
0.45 µm	*Serratia marcescens*	14756
0.65 µm	*Lactobacillus brevis*	

To determine the pore diameter, physical methods such as porosimeter (mercury, liquid-liquid porosimeter and Bubble Point Test) are also used, but a certain form of the pores (such as cylindrical or concatenated spherical holes) is assumed. Such methods are used for membranes whose pore geometry does not match the ideal, and we get "nominal" pore diameter, which characterizes the membrane, but does not necessarily reflect its actual filtration behavior and selectivity.

The selectivity is highly dependent on the separation process, the composition of the membrane and its electrochemical properties in addition to the pore size. With high selectivity, isotopes can be enriched (uranium enrichment) in nuclear engineering or industrial gases like nitrogen can be recovered (gas separation). Ideally, even racemics can be enriched with a suitable membrane.

When choosing membranes selectivity has priority over a high permeability, as low flows can easily be offset by increasing the filter surface with a modular structure. In gas phase filtration different deposition mechanisms are operative, so that particles having sizes below the pore size of the membrane can be retained as well.

7.7 MEMBRANE CLASSIFICATION

Bio-Membrane is classified in two categories, synthetic membrane and natural membrane. synthetic membranes further classified in organic and inorganic membranes. Organic membrane sub classified polymeric membranes and inorganic membrane sub classified ceramic polymers.

7.8 SYNTHESIS OF BIOMASS MEMBRANE

7.8.1 The Composite Biomass Membrane

Green membrane or Bio-membrane synthesis is the solution to protected environments which have largely comprehensive performance. Biomass is used in the form of activated carbon nanoparticles, like using cellulose based biomass coconut shell, hazelnut shell, walnut shell, agricultural wastes of corn stalks etc. which improve surface hydrophilicity, larger pore size, more and lower surface roughness therefore, the separation and anti-fouling performance of membranes are also improved simultaneously.

7.8.2 Fabrication of Pure Biomass Based Membrane

A biomass-based membrane is a membrane made from organic materials such as plant fibers. These membranes are often used in water filtration and wastewater treatment applications. The fabrication of a pure biomass-based membrane is a complex process that involves a number of steps. The first step is to create a slurry of the organic materials.

This slurry is then cast onto a substrate, such as a glass or metal plate. The cast is then dried, and the resulting membrane is then subjected to a number of treatments, such as chemical or heat treatments, to improve its properties. One of the challenges in the fabrication of biomass-based membranes is to create a membrane with the desired properties.

7.9 EQUIPMENT AND INSTRUMENTS USED IN THE PROCESS

- Centrifuge.
- Casting Machine.
- Plane casting glass.
- Magnetic Stirrer.
- Glass ware: Beakers, measuring cylinders, flask etc.
- Oven.
- Mortar and pestle.

7.10 MEMBRANE CHARACTERIZATION

After casting and synthesis of membrane there is need to characterize the prepared membrane to know more details about membrane parameters, like pore size, functional groups, wettability, surface charge, etc. It is important to know membrane properties so we are able to remove and treat a particulate pollutant, which causes pollution in the environment. For characterization following different instruments are used:

- Scanning Electron Microscope (SEM).
- Transmission electron Microscope (TEM).
- Fourier Transform Infrared Spectroscopy (FTIR).
- Atomic force microscopy.
- Contact angle meter.
- Zeta potential (streaming potential).

- X-ray Diffraction (XRD).
- Liquid–Liquid Displacement Porosimetry (LLDP).

7.11 BIOMASS MEMBRANE APPLICATIONS

- **Water Treatment:** Water treatment is any process that improves the quality of water to make it more acceptable for a specific end-use. Membranes can be used to remove particulates from water by either size exclusion or charge separation. In size exclusion, the pores in the membrane are sized such that only particles smaller than the pores can pass through. The pores in the membrane are sized such that only water molecules can pass through, leaving dissolved contaminants behind.
- **Gas Separation:** Utilization of membranes in gas separation, like carbon dioxide (CO_2), Nitrogen oxides (NO_x), Sulphur oxides (SO_x), harmful gasses can be removed to protect the environment. Biomass Membrane gas separation more effective than commercial membrane.
- **Hemodialysis:** Membrane application in hemodialysis is a process of using a semipermeable membrane to remove waste products and excess fluids from the blood.

CHAPTER

Precipitation

8

Precipitation is a widely utilized and fundamental method in the field of bioseparations, serving as a cost-effective and efficient approach for the separation and purification of biomolecules such as proteins, nucleic acids, and polysaccharides. It involves the conversion of soluble components into an insoluble form by altering the conditions of the surrounding environment, which leads to their aggregation and subsequent removal from solution. The process is often employed during the early or intermediate stages of purification due to its scalability and simplicity. Precipitation plays a vital role in both laboratory-scale and industrial bioseparation processes and continues to be an area of ongoing research and development.

8.1 PRINCIPLE AND MECHANISM OF PRECIPITATION

The principle of precipitation is based on reducing the solubility of the target biomolecule in the solution so that it aggregates and forms a solid phase. This can be achieved by changing the chemical environment, such as altering pH, temperature, ionic strength, or by adding specific precipitants. Once the solubility threshold is crossed, molecules begin to interact more strongly with one another than with the solvent, leading to aggregation. As aggregates grow larger, they eventually become heavy enough to sediment or be filtered out. The process can be finely tuned to selectively precipitate desired molecules while leaving unwanted components in the solution.

In protein precipitation, the technique takes advantage of the fact that proteins have different solubility profiles under various physicochemical conditions. For example, at a certain pH known as the isoelectric point, a protein carries no net charge, and its solubility tends to be minimal, increasing the likelihood of precipitation. Similarly, the addition of salts, organic solvents, or polymers can compete with the protein for water molecules, effectively reducing protein solubility and facilitating precipitation.

8.2 TYPES OF PRECIPITATION METHODS

There are various types of precipitation methods, each exploiting a specific factor to induce biomolecule precipitation. One common method is salt-induced precipitation, often referred to as salting out. This technique typically uses neutral salts such as ammonium sulfate, which draw water molecules away from proteins, decreasing their solubility and causing them to aggregate. Ammonium sulfate precipitation is particularly popular due to its gentle effect on proteins and its ability to preserve biological activity during separation.

Organic solvent precipitation is another widely used technique, especially in the case of nucleic acid separation. Solvents such as ethanol and isopropanol reduce the dielectric constant of the solution and disrupt the hydration shells of biomolecules, causing them to precipitate. This method is especially effective for precipitating DNA and RNA, and is routinely used in molecular biology laboratories.

Isoelectric precipitation takes advantage of the fact that proteins have minimal solubility at their isoelectric point. By adjusting the pH of the solution to this specific value, electrostatic repulsion between protein molecules is minimized, promoting aggregation and precipitation. This technique is especially useful for selectively precipitating individual proteins from complex mixtures.

Polymer-induced precipitation utilizes high molecular weight compounds such as polyethylene glycol (PEG) to reduce the availability of water for protein hydration. PEG occupies significant space in the solution and promotes the exclusion of water from the protein surface, leading to aggregation. This method is especially valuable for its mild conditions and ability to concentrate and fractionate proteins without denaturation.

8.3 APPLICATIONS OF PRECIPITATION IN BIOSEPARATIONS

Precipitation is employed in a variety of applications within the bioseparation field. In the pharmaceutical and biotechnology industries, it is often used as a primary purification step for recombinant proteins, enzymes, and antibodies produced in microbial or mammalian cell cultures. This step helps reduce the volume of the solution and remove major contaminants, setting the stage for more refined downstream purification techniques such as chromatography.

In the field of molecular biology, precipitation methods are routinely used to isolate and purify nucleic acids. For example, ethanol or isopropanol precipitation of DNA and RNA is a standard step following enzymatic reactions or cell lysis. These methods enable researchers to concentrate nucleic acids and remove unwanted salts, enzymes, and other impurities.

Food and dairy industries also make use of precipitation techniques to isolate proteins such as casein from milk or plant-derived proteins from soybean and other legumes. This method is favored due to its low cost and compatibility with large-scale processing. In environmental biotechnology, precipitation is used for the removal of biomolecules or biosolids from wastewater, aiding in pollution control and resource recovery.

8.4 ADVANTAGES OF PRECIPITATION IN BIOSEPARATION PROCESSES

One of the major advantages of precipitation is its simplicity and ease of scale-up. Unlike more complex and expensive methods like chromatography, precipitation can be performed using relatively inexpensive reagents and basic equipment. This makes it an ideal choice for large-scale industrial operations where cost-efficiency is a primary concern.

Precipitation is also a relatively rapid process, especially when compared to techniques that require long column runs or multiple buffer exchanges. It can be used to concentrate large volumes of dilute solutions, thereby improving process efficiency and reducing downstream processing time and cost.

Another important advantage is the selectivity and versatility that precipitation can offer. By carefully controlling the process parameters such as pH, temperature, or the type and concentration of precipitant, it is possible to target specific classes of biomolecules or even individual components within a mixture. This selectivity makes it suitable for fractionation, where different proteins or biomolecules can be separated based on their precipitation profiles.

8.5 LIMITATIONS AND CHALLENGES

Despite its many advantages, precipitation is not without limitations. One significant drawback is the potential for incomplete separation, especially when multiple components in the mixture have similar solubility characteristics. This can result in co-precipitation of impurities, reducing the overall purity of the target product.

Additionally, precipitation conditions must be carefully optimized to prevent denaturation or loss of biological activity. Some precipitants, particularly organic solvents or strong acids, can cause irreversible changes to the structure of sensitive proteins, rendering them inactive. Furthermore, the recovery and redissolution of the precipitated product can sometimes be challenging, especially if the aggregated material forms tight or insoluble clumps.

Scalability, while generally favorable, can also pose difficulties in terms of uniform mixing and control over precipitation kinetics in large reactors. Poor mixing or localized supersaturation can lead to heterogeneous precipitation and reduced reproducibility.

8.6 RECENT DEVELOPMENTS AND FUTURE PERSPECTIVES

The field of bioseparations is constantly evolving, and new approaches to precipitation are being explored to enhance its performance and address current limitations. One promising direction is the use of smart polymers and stimuli-responsive precipitants that can change their solubility in response to external triggers such as temperature, pH, or light. These materials offer the possibility of reversible and highly controlled precipitation processes.

Advances in process analytical technologies (PAT) and in-line monitoring tools are also improving the ability to control and optimize precipitation steps. Real-time monitoring of turbidity, pH, and conductivity enables better process understanding and control, leading to more consistent and high-quality outcomes.

Integration of precipitation with other downstream techniques, such as membrane filtration or chromatography, is also being explored to develop hybrid purification schemes. These integrated systems aim to combine the strengths of multiple methods to achieve higher purity, yield, and efficiency.

8.7 CONCLUSION

Precipitation remains a vital and highly practical method in the realm of bioseparations. Its ability to efficiently concentrate and partially purify biomolecules makes it an indispensable tool across various industries. Although it faces certain challenges, ongoing research and technological advancements are expanding its capabilities and improving its reliability. As the demand for high-purity bioproducts continues to grow, precipitation techniques, especially when combined with modern innovations, will continue to play a key role in meeting these needs and advancing the field of bioprocessing.

CHAPTER 9

Chromatography: Principles and Practice

In chemical analysis, chromatography is a laboratory technique for the separation of a mixture into its components. The mixture is dissolved in a fluid solvent (gas or liquid) called the mobile phase, which carries it through a system (a column, a capillary tube, a plate, or a sheet) on which a material called the stationary phase is fixed. Because the different constituents of the mixture tend to have different affinities for the stationary phase and are retained for different lengths of time depending on their interactions with its surface sites, the constituents travel at different apparent velocities in the mobile fluid, causing them to separate. The separation is based on the differential partitioning between the mobile and the stationary phases. Subtle differences in a compound's partition coefficient result in differential retention on the stationary phase and thus affect the separation.

Chromatography may be preparative or analytical. The purpose of preparative chromatography is to separate the components of a mixture for later use, and is thus a form of purification. This process is associated with higher costs due to its mode of production. Analytical chromatography is done normally with smaller amounts of material and is for establishing the presence or measuring the relative proportions of analytes in a mixture. The two types are not mutually exclusive.

9.1 HISTORY OF CHROMATOGRAPHY

The history of chromatography spans from the mid-19th century to the 21st. Chromatography, literally "color writing", was used—and named— in the first decade of the 20th century, primarily for the separation of plant pigments such as chlorophyll (which is green) and carotenoids (which are orange and yellow). New forms of chromatography developed in the 1930s and 1940s made the technique useful for a wide range of separation processes and chemical analysis tasks, especially in biochemistry.

9.1.1 Precursors

The earliest use of chromatography—passing a mixture through an inert material to create separation of the solution components based on differential adsorption—is sometimes attributed to German chemist Friedlieb Ferdinand Runge, who in

1855 described the use of paper to analyze dyes. Runge dropped spots of different inorganic chemicals onto circles of filter paper already impregnated with another chemical, and reactions between the different chemicals created unique color patterns. According to historical analysis of L. S. Ettre, however, Runge's work had "nothing to do with chromatography" (and instead should be considered a precursor of chemical spot tests such as the Schiff test).

In the 1860s, Christian Friedrich Schönbein and his student Friedrich Goppelsroeder published the first attempts to study the different rates at which different substances move through filter paper. Schönbein, who thought capillary action (rather than adsorption) was responsible for the movement, called the technique capillary analysis, and Goppelsroeder spent much of his career using capillary analysis to test the movement rates of a wide variety of substances. Unlike modern paper chromatography, capillary analysis used reservoirs of the substance being analyzed, creating overlapping zones of the solution components rather than separate points or bands.

Work on capillary analysis continued, but without much technical development, well into the 20th century. The first significant advances over Goppelsroeder's methods came with the work of Raphael E. Liesegang: in 1927, he placed filter strips in closed containers with atmospheres saturated by solvents, and in 1943 he began using discrete spots of sample adsorbed to filter paper, dipped in pure solvent to achieve separation. This method, essentially identical to modern paper chromatography, was published just before the independent—and far more influential—work of Archer Martin and his collaborators that inaugurated the widespread use of paper chromatography.

In 1897, the American chemist David Talbot Day (1859–1915), then serving with the U.S. Geological Survey, observed that crude petroleum generated bands of color as it seeped upwards through finely divided clay or limestone. In 1900, he reported his findings at the First International Petroleum Congress in Paris, where they created a sensation.

9.1.2 Tsvet and Column Chromatography

The first true chromatography is usually attributed to the Russian-Italian botanist Mikhail Tsvet. Tsvet applied his observations with filter paper extraction to the new methods of column fractionation that had been developed in the 1890s for separating the components of petroleum. He used a liquid-adsorption column containing calcium carbonate to separate yellow, orange, and green plant pigments (what are known today as xanthophylls, carotenes, and chlorophylls, respectively). The method was described on December 30, 1901, at the 11th Congress of Naturalists and Doctors in Saint Petersburg. The first printed description was in 1903, in the

Proceedings of the Warsaw Society of Naturalists, section of biology. He first used the term chromatography in print in 1906 in his two papers about chlorophyll in the German botanical journal, Berichte der Deutschen Botanischen Gesellschaft. In 1907 he demonstrated his chromatograph for the German Botanical Society. Mikhail's surname means "color" in Russian, so there is the possibility that his naming the procedure chromatography (literally "color writing") was a way that he could make sure that he, a commoner in Tsarist Russia, could be immortalized.

In a 1903 lecture (published in 1905), Tsvet also described using filter paper to approximate the properties of living plant fibers in his experiments on plant pigments—a precursor to paper chromatography. He found that he could extract some pigments (such as orange carotenes and yellow xanthophylls) from leaves with non-polar solvents, but others (such as chlorophyll) required polar solvents. He reasoned that chlorophyll was held to the plant tissue by adsorption, and that stronger solvents were necessary to overcome the adsorption. To test this, he applied dissolved pigments to filter paper, allowed the solvent to evaporate, then applied different solvents to see which could extract the pigments from the filter paper. He found the same pattern as from leaf extractions: carotene could be extracted from filter paper using non-polar solvents, but chlorophyll required polar solvents. Tsvet's work saw little use until the 1930s.

9.1.3 Martin and Synge and Partition Chromatography

Chromatography methods changed little after Tsvet's work until the explosion of mid-20th century research in new techniques, particularly thanks to the work of Archer John Porter Martin and Richard Laurence Millington Synge. By "the marrying of two techniques, that of chromatography and that of countercurrent solvent extraction", Martin and Synge developed partition chromatography to separate chemicals with only slight differences in partition coefficients between two liquid solvents. Martin, who had previously been working in vitamin chemistry (including attempts to purify vitamin E), began collaborating with Synge in 1938, brought his experience with equipment design to Synge's project of separating amino acids. After unsuccessful experiments with complex countercurrent extraction machines and liquid-liquid chromatography methods where the liquids move in opposite directions, Martin hit on the idea of using silica gel in columns to hold water stationary while an organic solvent flows through the column. Martin and Synge demonstrated the potential of the methods by separating amino acids marked in the column by the addition of methyl red. In a series of publications beginning in 1941, they described increasingly powerful methods of separating amino acids and other organic chemicals.

In pursuit of better and easier methods of identifying the amino acid constituents of peptides, Martin and Synge turned to other chromatography media as well. A short abstract in 1943 followed by a detailed article in 1944 described the use of filter paper as the stationary phase for performing chromatography on amino acids: paper chromatography. By 1947, Martin, Synge and their collaborators had applied this method (along with Fred Sanger's reagent for identifying N-terminal residues) to determine the pentapeptide sequence of Gramicidin S. These and related paper chromatography methods were also foundational to Fred Sanger's effort to determine the amino acid sequence of insulin.

9.1.4 Refining the Techniques

Martin, in collaboration with Anthony T. James, went on to develop gas chromatography (GC; the principles of which Martin and Synge had predicted in their landmark 1941 paper) beginning in 1949. In 1952, during his lecture for the Nobel Prize in Chemistry (shared with Synge, for their earlier chromatography work), Martin announced the successful separation of a wide variety of natural compounds by gas chromatography. Previously, Erika Cremer had laid the theoretical basis of GC in 1944 and Austrian chemist Fritz Prior, under the direction of Erika Cremer, constructed in 1947 the first prototype of a gas chromatograph and achieved separating oxygen and carbon dioxide, in 1947 during his Ph.D. research.

The ease and efficiency of gas chromatography for separating organic chemicals spurred the rapid adoption of the method, as well as the rapid development of new detection methods for analyzing the output. The thermal conductivity detector, described in 1954 by N. H. Ray, was the foundation for several other methods: the flame ionization detector was described by J. Harley, W. Nel, and V. Pretorius in 1958, and James Lovelock introduced the electron capture detector that year as well. Others introduced mass spectrometers to gas chromatography in the late 1950s.

The work of Martin and Synge also set the stage for high performance liquid chromatography, suggesting that small sorbent particles and pressure could produce fast liquid chromatography techniques. This became widely practical by the late 1960s (and the method was used to separate amino acids as early as 1960).

9.1.5 Thin Layer Chromatography

The first developments in thin layer chromatography occurred in the 1940s, and techniques advanced rapidly in the 1950s after the introduction of relatively large plates and relatively stable materials for sorbent layers.

9.1.6 Later Developments

In 1987 Pedro Cuatrecasas and Meir Wilchek were awarded the Wolf Prize in Medicine for the invention and development of affinity chromatography and its applications to biomedical sciences.

9.2 TERMS

- **Analyte:** The substance to be separated during chromatography. It is also normally what is needed from the mixture.
- **Analytical Chromatography:** The use of chromatography to determine the existence and possibly also the concentration of analyte(s) in a sample.
- **Bonded Phase:** A stationary phase that is covalently bonded to the support particles or to the inside wall of the column tubing.
- **Chromatogram:** The visual output of the chromatograph. In the case of an optimal separation, different peaks or patterns on the chromatogram correspond to different components of the separated mixture.
- **Chromatograph:** An instrument that enables a sophisticated separation, e.g. gas chromatographic or liquid chromatographic separation.
- **Chromatography:** A physical method of separation that distributes components to separate between two phases, one stationary (stationary phase), the other (the mobile phase) moving in a definite direction.
- **Eluent (Sometimes Spelled Eluant):** The solvent or solvent fixure used in elution chromatography and is synonymous with mobile phase.
- **Eluate:** The mixture of solute and solvent exiting the column.
- **Effluent:** The stream flowing out of a chromatographic column. In practise, it is used synonymously with eluate, but the term more precisely refers to the stream independent of separation taking place.
- **Eluite:** A more precise term for solute or analyte. It is a sample component leaving the chromatographic column.
- **Eluotropic Series:** A list of solvents ranked according to their eluting power.
- **Immobilized Phase:** A stationary phase that is immobilized on the support particles, or on the inner wall of the column tubing.
- **Mobile Phase:** The phase that moves in a definite direction. It may be a liquid (LC and capillary electrochromatography, CEC), a gas (GC), or a supercritical fluid (supercritical-fluid chromatography, SFC). The mobile phase consists of the sample being separated/analyzed and the solvent that moves the sample through the column. In the case of HPLC the mobile phase consists of a non-polar solvent(s) such as hexane in normal phase or a polar solvent such as methanol in reverse phase chromatography and the sample being separated. The mobile phase

moves through the chromatography column (the stationary phase) where the sample interacts with the stationary phase and is separated.

- **Preparative Chromatography:** The use of chromatography to purify sufficient quantities of a substance for further use, rather than analysis.
- **Retention Time:** The characteristic time it takes for a particular analyte to pass through the system (from the column inlet to the detector) under set conditions.
- **Sample:** The matter analyzed in chromatography. It may consist of a single component or it may be a mixture of components. When the sample is treated in the course of an analysis, the phase or the phases containing the analytes of interest is/are referred to as the sample whereas everything out of interest separated from the sample before or in the course of the analysis is referred to as waste.
- **Solute:** The sample components in partition chromatography.
- **Solvent:** Any substance capable of solubilizing another substance, and especially the liquid mobile phase in liquid chromatography.
- **Stationary Phase:** The substance fixed in place for the chromatography procedure. Examples include the silica layer in thin-layer chromatography
- **Detector:** The instrument used for qualitative and quantitative detection of analytes after separation.

Chromatography is based on the concept of partition coefficient. Any solute partitions between two immiscible solvents. When one make one solvent immobile (by adsorption on a solid support matrix) and another mobile it results in most common applications of chromatography. If the matrix support, or stationary phase, is polar (e.g., cellulose, silica etc.) it is forward phase chromatography. Otherwise this technique is known as reversed phase, where a non-polar stationary phase (e.g., non-polar derivative of C-18) is used.

9.3 COLUMN CHROMATOGRAPHY

Column chromatography in chemistry is a chromatography method used to isolate a single chemical compound from a mixture. Chromatography is able to separate substances based on differential absorption of compounds to the adsorbent; compounds move through the column at different rates, allowing them to be separated into fractions. The technique is widely applicable, as many different adsorbents (normal phase, reversed phase, or otherwise) can be used with a wide range of solvents. The technique can be used on scales from micrograms up to kilograms. The main advantage of column chromatography is the relatively low cost and disposability of the stationary phase used in the process. The latter prevents cross-contamination and stationary phase degradation due to recycling. Column chromatography can be done using gravity to move the solvent, or using compressed gas to push the solvent through the column. A thin-layer chromatography can show how a mixture of compounds will behave when purified by column chromatography. The separation is first optimised using thin-layer chromatography before performing column chromatography.

9.3.1 Column Preparation

A column is prepared by packing a solid adsorbent into a cylindrical glass or plastic tube. The size will depend on the amount of compound being isolated. The base of the tube contains a filter, either a cotton or glass wool plug, or glass frit to hold the solid phase in place. A solvent reservoir may be attached at the top of the column.

Two methods are generally used to prepare a column: the dry method and the wet method. For the dry method, the column is first filled with dry stationary phase powder, followed by the addition of mobile phase, which is flushed through the column until it is completely wet, and from this point is never allowed to run dry. For the wet method, a slurry is prepared of the eluent with the stationary phase powder and then carefully poured into the column. The top of the silica should be flat, and the top of the silica can be protected by a layer of sand. Eluent is slowly passed through the column to advance the organic material.

The individual components are retained by the stationary phase differently and separate from each other while they are running at different speeds through the column with the eluent. At the end of the column they elute one at a time. During the entire chromatography process the eluent is collected in a series of fractions. Fractions can be collected automatically by means of fraction collectors. The productivity of chromatography can be increased by running several columns at a time. In this case multi stream collectors are used. The composition of the eluent flow can be monitored and each fraction is analyzed for dissolved compounds, e.g. by analytical chromatography, UV absorption spectra, or fluorescence. Colored compounds (or fluorescent compounds with the aid of a UV lamp) can be seen through the glass wall as moving bands.

9.3.2 Stationary Phase

The stationary phase or adsorbent in column chromatography is a solid. The most common stationary phase for column chromatography is silica gel, the next most common being alumina. Cellulose powder has often been used in the past. A wide range of stationary phases are available in order to perform ion exchange chromatography, reversed-phase chromatography (RP), affinity chromatography or expanded bed adsorption (EBA). The stationary phases are usually finely ground powders or gels and/or are microporous for an increased surface, though in EBA a fluidized bed is used. There is an important ratio between the stationary phase weight and the dry weight of the analyte mixture that can be applied onto the column. For silica column chromatography, this ratio lies within 20:1 to 100:1, depending on how close to each other the analyte components are being eluted.

9.3.3 Mobile Phase (Eluent)

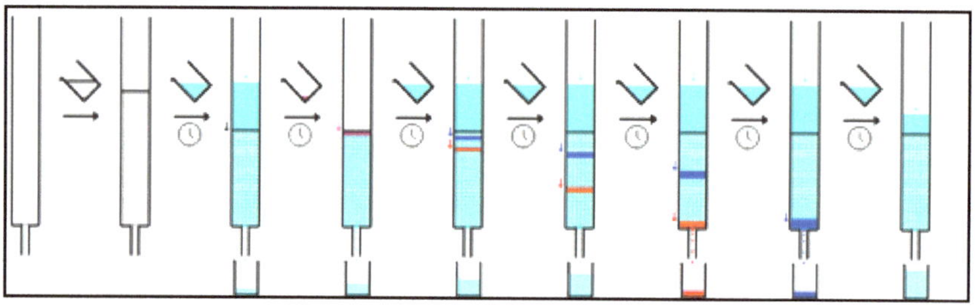

Fig: Column Chromatography Proceeds by a Series of Steps

The mobile phase or eluent is a solvent or a mixture of solvents used to move the compounds through the column. It is chosen so that the retention factor value of the compound of interest is roughly around 0.2 - 0.3 in order to minimize the time and the amount of eluent to run the chromatography.

The eluent has also been chosen so that the different compounds can be separated effectively. The eluent is optimized in small scale pretests, often using thin layer chromatography (TLC) with the same stationary phase, using solvents of different polarity until a suitable solvent system is found. Common mobile phase solvents, in order of increasing polarity, include hexane, dichloromethane, ethyl acetate, acetone, and methanol.

A common solvent system is a mixture of hexane and ethyl acetate, with proportions adjusted until the target compound has a retention factor of 0.2 - 0.3. Contrary to common misconception, methanol alone can be used as an eluent for highly polar compounds, and does not dissolve silica gel.

There is an optimum flow rate for each particular separation. A faster flow rate of the eluent minimizes the time required to run a column and thereby minimizes diffusion, resulting in a better separation. However, the maximum flow rate is limited because a finite time is required for the analyte to equilibrate between the stationary phase and mobile phase.

A simple laboratory column runs by gravity flow. The flow rate of such a column can be increased by extending the fresh eluent filled column above the top of the stationary phase or decreased by the tap controls. Faster flow rates can be achieved by using a pump or by using compressed gas (e.g. air, nitrogen, or argon) to push the solvent through the column (flash column chromatography).

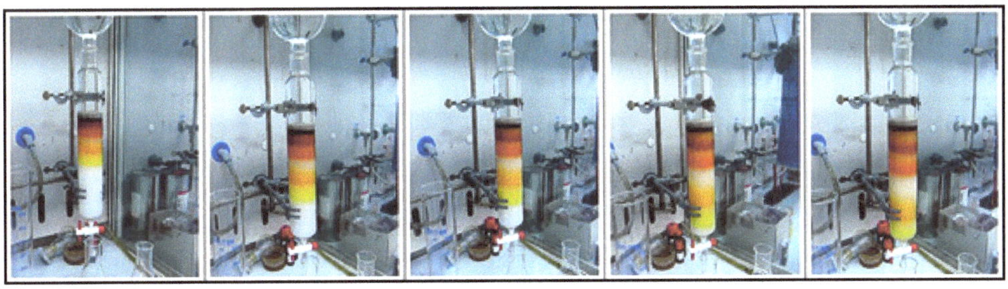

Fig: Photographic Sequence of a Column Chromatography

The particle size of the stationary phase is generally finer in flash column chromatography than in gravity column chromatography. For example, one of the most widely used silica gel grades in the former technique is mesh 230 – 400 (40 – 63 μm), while the latter technique typically requires mesh 70 – 230 (63 – 200 μm) silica gel.

A spreadsheet that assists in the successful development of flash columns has been developed. The spreadsheet estimates the retention volume and band volume of analytes, the fraction numbers expected to contain each analyte, and the resolution between adjacent peaks. This information allows users to select optimal parameters for preparative-scale separations before the flash column itself is attempted.

9.3.4 Automated Systems

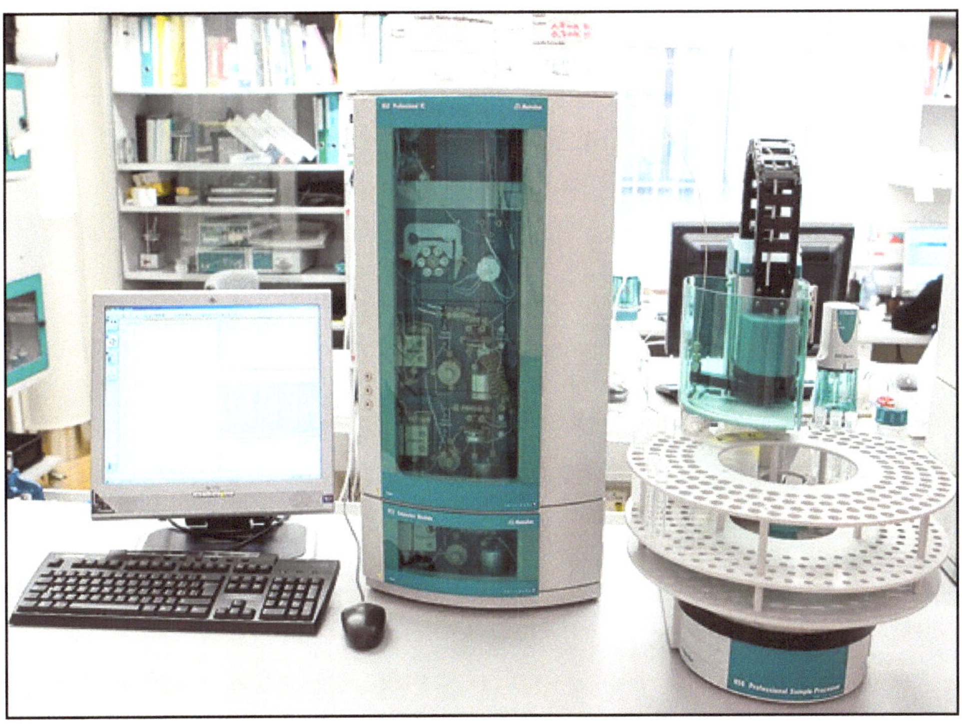

Fig: An Automated Ion Chromatography System

Column chromatography is an extremely time-consuming stage in any lab and can quickly become the bottleneck for any process lab. Many manufacturers like Biotage, Buchi, Interchim and Teledyne Isco have developed automated flash chromatography systems (typically referred to as LPLC, low pressure liquid chromatography, around 350–525 kPa or 50.8–76.1 psi) that minimize human involvement in the purification process. Automated systems will include components normally found on more expensive high performance liquid chromatography (HPLC) systems such as a gradient pump, sample injection ports, a UV detector and a fraction collector to collect the eluent. Typically these automated systems can separate samples from a few milligrams up to an industrial many kilogram scale and offer a much cheaper and quicker solution to doing multiple injections on prep-HPLC systems.

The resolution (or the ability to separate a mixture) on an LPLC system will always be lower compared to HPLC, as the packing material in an HPLC column can be much smaller, typically only 5 micrometre thus increasing stationary phase surface area, increasing surface interactions and giving better separation. However, the use of this small packing media causes the high back pressure and is why it is termed high pressure liquid chromatography. The LPLC columns are typically packed with silica of around 50 micrometres, thus reducing back pressure and resolution, but it also removes the need for expensive high pressure pumps. Manufacturers are now starting to move into higher pressure flash chromatography systems and have termed these as medium pressure liquid chromatography (MPLC) systems which operate above 1 MPa (150 psi).

9.3.5 Column Chromatogram Resolution Calculation

Typically, column chromatography is set up with peristaltic pumps, flowing buffers and the solution sample through the top of the column. The solutions and buffers pass through the column where a fraction collector at the end of the column setup collects the eluted samples. Prior to the fraction collection, the samples that are eluted from the column pass through a detector such as a spectrophotometer or mass spectrometer so that the concentration of the separated samples in the sample solution mixture can be determined.

For example, if you were to separate two different proteins with different binding capacities to the column from a solution sample, a good type of detector would be a spectrophotometer using a wavelength of 280 nm. The higher the concentration of protein that passes through the eluted solution through the column, the higher the absorbance of that wavelength.

Because the column chromatography has a constant flow of eluted solution passing through the detector at varying concentrations, the detector must plot the concentration of the eluted sample over a course of time. This plot of sample concentration versus time is called a chromatogram.

The ultimate goal of chromatography is to separate different components from a solution mixture. The resolution expresses the extent of separation between the components from the mixture. The higher the resolution of the chromatogram, the better the extent of separation of the samples the column gives. This data is a good way of determining the column's separation properties of that particular sample. The resolution can be calculated from the chromatogram. The separate curves in the diagram represent different sample elution concentration profiles over time based on their affinity to the column resin. To calculate resolution, the retention time and curve width are required. Retention time is the time from the start of signal detection by the detector to the peak height of the elution concentration profile of each different sample.

Curve width is the width of the concentration profile curve of the different samples in the chromatogram in units of time. A simplified method of calculating chromatogram resolution is to use the plate model. The plate model assumes that the column can be divided into a certain number of sections, or plates and the mass balance can be calculated for each individual plate. This approach approximates a typical chromatogram curve as a Gaussian distribution curve. By doing this, the curve width is estimated as 4 times the standard deviation of the curve, 4σ. The retention time is the time from the start of signal detection to the time of the peak height of the Gaussian curve.

The resolution, plate number, and plate height of the column plate model can be calculated using the equations:

Resolution (R_s):

$$R_s = 2(t_{RB} - t_{RA})/(w_B + w_A),$$

where:

- t_{RB} = retention time of solute B
- t_{RA} = retention time of solute A
- w_B = Gaussian curve width of solute B
- w_A = Gaussian curve width of solute A

Plate Number (N):

- $N = (t_R)^2/(w/4)^2$

Plate Height (H):

- $H = L/N$

where L is the length of the column.

9.3.6 Column Adsorption Equilibrium

For an adsorption column, the column resin (the stationary phase) is composed of microbeads. Even smaller particles such as proteins, carbohydrates, metal ions, or other chemical compounds are conjugated onto the microbeads. Each binding particle that is attached to the microbead can be assumed to bind in a 1:1 ratio with the solute sample sent through the column that needs to be purified or separated.

Binding between the target molecule to be separated and the binding molecule on the column beads can be modeled using a simple equilibrium reaction $K_{eq} = [CS]/([C][S])$ where K_{eq} is the equilibrium constant, [C] and [S] are the concentrations of the target molecule and the binding molecule on the column resin, respectively. [CS] is the concentration of the complex of the target molecule bound to the column resin.

Using this as a basis, three different isotherms can be used to describe the binding dynamics of a column chromatography: linear, Langmuir, and Freundlich.

The linear isotherm occurs when the solute concentration needed to be purified is very small relative to the binding molecule. Thus, the equilibrium can be defined as:

$$[CS] = K_{eq}[C].$$

For industrial scale uses, the total binding molecules on the column resin beads must be factored in because unoccupied sites must be taken into account. The Langmuir isotherm and Freundlich isotherm are useful in describing this equilibrium. The Langmuir isotherm is given by:

$[CS] = (K_{eq}S_{tot}[C])/(1 + K_{eq}[C])$, where S_{tot} is the total binding molecules on the beads.

The Freundlich isotherm is given by:

$$[CS] = K_{eq}[C]^{1/n}$$

The Freundlich isotherm is used when the column can bind to many different samples in the solution that needs to be purified. Because the many different samples have different binding constants to the beads, there are many different K_{eq}s. Therefore, the Langmuir isotherm is not a good model for binding in this case.

9.4 PLANAR CHROMATOGRAPHY

Planar chromatography is a separation technique in which the stationary phase is present as or on a plane. The plane can be a paper, serving as such or impregnated by a substance as the stationary bed (paper chromatography) or a layer of solid particles spread on a support such as a glass plate (thin-layer chromatography). Different compounds in the sample mixture travel different distances according to how strongly they interact with the stationary phase as compared to the mobile phase. The specific Retention factor (R_f) of each chemical can be used to aid in the identification of an unknown substance.

9.4.1 Paper Chromatography

Paper chromatography is an analytical method used to separate colored chemicals or substances. It can also be used for colorless chemicals that can be located by a stain or other visualisation method after separation. It is now primarily used as a teaching tool, having been replaced in the laboratory by other chromatography methods such as thin-layer chromatography (TLC).

This analytic method has three components, a mobile phase, stationary phase and a support medium (the paper). The mobile phase is generally a non-polar organic solvent in which the sample is dissolved. The stationary phase consists of (polar) water molecules that were incorporated into the paper when it was manufactured. The mobile phase travels up the stationary phase by capillary action, carrying the sample with it. The difference between TLC and paper chromatography is that the stationary phase in TLC is a layer of adsorbent (usually silica gel, or aluminium oxide), and the stationary phase in paper chromatography is less absorbent paper.

A paper chromatography variant, two-dimensional chromatography, involves using two solvents and rotating the paper 90° in between. This is useful for separating complex mixtures of compounds having similar polarity, for example, amino acids.

Rf Value, Solutes, and Solvents

The retention factor (R_f) may be defined as the ratio of the distance travelled by the solute to the distance travelled by the solvent. It is used in chromatography to quantify the amount of retardation of a sample in a stationary phase relative to a mobile phase. R_f values are usually expressed as a fraction of two decimal places.

- If R_f value of a solution is zero, the solute remains in the stationary phase and thus it is immobile.
- If R_f value = 1 then the solute has no affinity for the stationary phase and travels with the solvent front.

For example, if a compound travels 9.9 cm and the solvent front travels 12.7 cm, the R_f value = (9.9/12.7) = 0.779 or 0.78. R_f value depends on temperature and the solvent used in experiment, so several solvents offer several R_f values for the same mixture of compound. A solvent in chromatography is the liquid the paper is placed in, and the solute is the ink which is being separated.

Pigments and Polarity

Paper chromatography is one method for testing the purity of compounds and identifying substances. Paper chromatography is a useful technique because it is relatively quick and requires only small quantities of material. Separations in paper chromatography involve the principle of partition. In paper chromatography, substances are distributed between a stationary phase and a mobile phase. The stationary phase is the water trapped between the cellulose fibers of the paper. The mobile phase is a developing solution that travels up the stationary phase, carrying the samples with it. Components of the sample will separate readily according to how strongly they adsorb onto the stationary phase versus how readily they dissolve in the mobile phase.

When a colored chemical sample is placed on a filter paper, the colors separate from the sample by placing one end of the paper in a solvent. The solvent diffuses up the paper, dissolving the various molecules in the sample according to the polarities of the molecules and the solvent.

If the sample contains more than one color, that means it must have more than one kind of molecule. Because of the different chemical structures of each kind of molecule, the chances are very high that each molecule will have at least a slightly different polarity, giving each molecule a different solubility in the solvent. The unequal solubility causes the various color molecules to leave solution at different places as the solvent continues to move up the paper. The more soluble a molecule is, the higher it will migrate up the paper. If a chemical is very non-polar it will not dissolve at all in a very polar solvent. This is the same for a very polar chemical and a very non-polar solvent.

When using water (a very polar substance) as a solvent, the more polar the color, the higher it will rise on the papers.

Types

- **Descending:** Development of the chromatogram is done by allowing the solvent to travel down the paper. Here, the mobile phase is placed in a solvent holder at the top.
- **Ascending:** Here the solvent travels up the chromatographic paper. Both descending and ascending paper chromatography are used for the separation of organic and inorganic substances. The sample and solvent move upward.
- **The Ascending and Descending Method:** The upper part of ascending chromatography can be folded over a rod in order to allow the paper to become descending after crossing the rod.
- **Circular Chromatography:** A circular filter paper is taken and the sample is deposited at the center of the paper. After drying the spot, the filter paper is tied horizontally on a Petri dish containing solvent, so that the wick of the paper is dipped in the solvent. The solvent rises

through the wick and the components are separated into concentric rings.

- **Two-Dimensional:** In this technique a square or rectangular paper is used. Here the sample is applied to one of the corners and development is performed at a right angle to the direction of the first run.

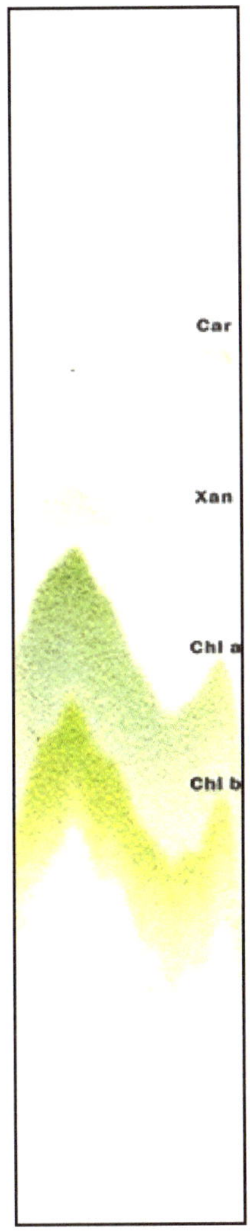

Fig: Taxus Baccata Paper Chromatography

9.4.2 Thin-Layer Chromatography (TLC)

Thin-layer chromatography (TLC) is a chromatography technique that separates components in non-volatile mixtures. It is performed on a TLC plate made up of a non-reactive solid coated with a thin layer of adsorbent material. This is called the stationary phase. The sample is deposited on the plate, which is eluted with a solvent or solvent mixture known as the mobile phase (or eluent). This solvent then moves up the plate via capillary action. As with all chromatography, some compounds are more attracted to the mobile phase, while others are more attracted to the stationary phase. Therefore, different compounds move up the TLC plate at different speeds and become separated. To visualize colourless compounds, the plate is viewed under UV light or is stained. Testing different stationary and mobile phases is often necessary to obtain well-defined and separated spots. TLC is quick, simple, and gives high sensitivity for a relatively low cost. It can monitor reaction progress, identify compounds in a mixture, determine purity, or purify small amounts of compound.

Procedure

The process for TLC is similar to paper chromatography but provides faster runs, better separations, and the choice between different stationary phases. Plates can be labelled before or after the chromatography process with a pencil or other implement that will not interfere with the process.

There are four main stages to running a thin-layer chromatography plate:

- **Plate Preparation:** Using a capillary tube, a small amount of a concentrated solution of the sample is deposited near the bottom edge of a TLC plate. The solvent is allowed to completely evaporate before the next step. A vacuum chamber may be necessary for non-volatile solvents. To make sure there is sufficient compound to obtain a visible result, the spotting procedure can be repeated. Depending on the application, multiple different samples may be placed in a row the same distance from the bottom edge; each sample will move up the plate in its own "lane."

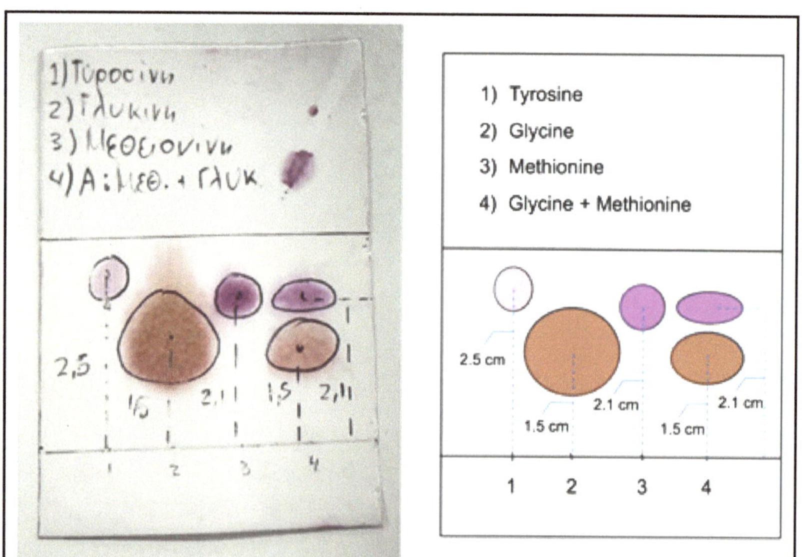

Fig: TLC of Three Amino Acids and a Sample (Left) with an English Translation (Right)

- **Development Chamber Preparation:** The development solvent or solvent mixture is placed into a transparent container (separation/ development chamber) to a depth of less than 1 centimetre. A strip of filter paper (aka "wick") is also placed along the container wall. This filter paper should touch the solvent and almost reach the top of the container. The container is covered with a lid and the solvent vapors are allowed to saturate the atmosphere of the container. Failure to do so results in poor separation and non-reproducible results.
- **Development:** The TLC plate is placed in the container such that the sample spot(s) are not submerged into the mobile phase. The container is covered to prevent solvent evaporation. The solvent migrates up the plate by capillary action, meets the sample mixture, and carries it up the plate (elutes the sample). The plate is removed from the container before the solvent reaches the top of the plate; otherwise, the results will be misleading. The solvent front, the highest mark the solvent has travelled along the plate, is marked.
- **Visualization:** The solvent evaporates from the plate. Visualization methods include UV light, staining, and many more.

Separation Process and Principle

The separation of compounds is due to the differences in their attraction to the stationary phase and because of differences in solubility in the solvent. As a result, the compounds and the mobile phase compete for binding sites on the stationary phase. Different compounds in the sample mixture travel at different rates due to the differences in their partition coefficients. Different solvents, or different solvent mixtures, gives different separation. The retardation factor (R_f), or retention factor, quantifies the results. It is the distance traveled by a given substance divided by the distance traveled by the mobile phase.

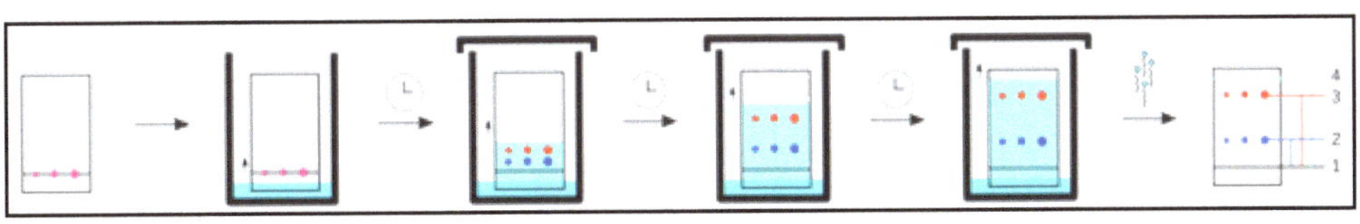

Fig: Development of a TLC Plate. Spots that Appear Purple Separate into Red Spots and Blue Spots

In normal-phase TLC, the stationary phase is polar. Silica gel is very common in normal-phase TLC. More polar compounds in a sample mixture interact more strongly with the polar stationary phase. As a result, more-polar compounds move less (resulting in smaller R_f) while less-polar compounds move higher up the plate (higher R_f). A more-polar mobile phase also binds more strongly to the plate, competing more with the compound for binding sites; a more-polar mobile phase also dissolves polar compounds more. As such, all compounds on the TLC plate move higher up the plate in polar solvent mixtures. "Strong" solvents move compounds higher up the plate, whereas "weak" solvents move them less.

If the stationary phase is non-polar, like C18-functionalized silica plates, it is called reverse-phase TLC. In this case, non-polar compounds move less and polar compounds move more. The solvent mixture will also be much more polar than in normal-phase TLC.

Solvent Choice

An eluotropic series, which orders solvents by how much they move compounds, can help in selecting a mobile phase. Solvents are also divided into solvent selectivity groups. Using solvents with different elution strengths or different selectivity groups can often give very different results. While single-solvent mobile phases can sometimes give good separation, some cases may require solvent mixtures.

In normal-phase TLC, the most common solvent mixtures include ethyl acetate/hexanes (EtOAc/Hex) for less-polar compounds and methanol/dichloromethane (MeOH/DCM) for more polar compounds. Different solvent mixtures and solvent ratios can help give better separation. In reverse-phase TLC, solvent mixtures are typically water with a less-polar solvent: Typical choices are water with tetrahydrofuran (THF), acetonitrile (ACN), or methanol.

Analysis

As the chemicals being separated may be colourless, several methods exist to visualise the spots:

- Placing the plate under blacklight (366 nm light) makes fluorescent compounds glow
- TLC plates containing a small amount of fluorescent compound (usually manganese-activated zinc silicate) in the adsorbent layer allow for visualisation of some compounds under UV-C light (254 nm). The adsorbent layer will fluoresce light-green, while spots containing compounds that absorb UV-C light will not.
- Placing the plate in a container filled with iodine vapours temporarily stains the spots. They typically become a yellow or brown colour.
- The TLC plate can either be dipped in or sprayed with a stain and sometimes heated depending on the stain used. Many stains exist for a large range of chemical moieties but some examples include:
  - Potassium permanganate (no heating, for oxidisable groups);
  - Ninhydrin (heating, amines and amino-acids);

  - Acidic vanillin (heating, general reagent); and
  - Phosphomolybdic acid (no heating, general reagent).
- In the case of lipids, the chromatogram may be transferred to a polyvinylidene fluoride membrane and then subjected to further analysis, for example, mass spectrometry. This technique is known as far-eastern blot.

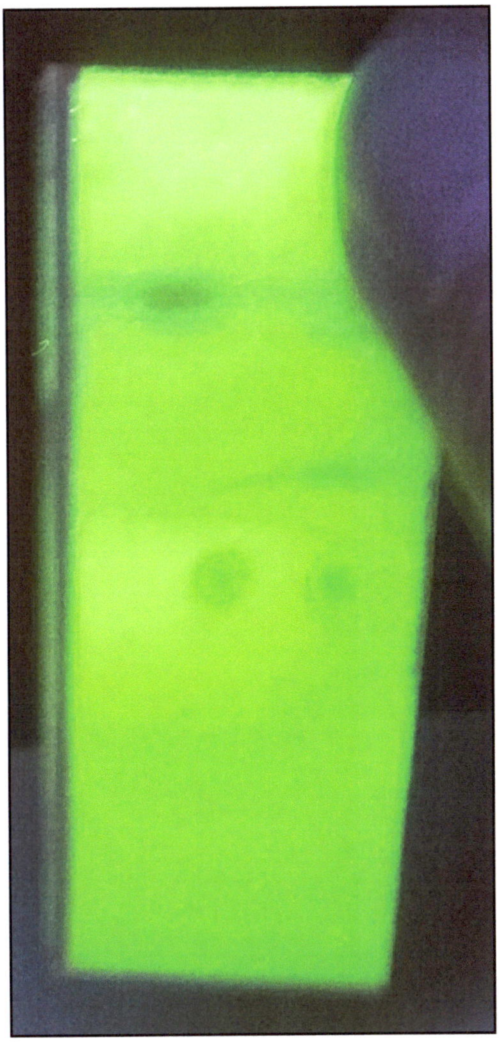

Fig: TLC Plate Visualised with UV-Light

Plate Production

TLC plates are usually commercially available, with standard particle size ranges to improve reproducibility. They are prepared by mixing the adsorbent, such as silica gel, with a small amount of inert binder like calcium sulfate (gypsum) and water. This mixture is spread as a thick slurry on an unreactive carrier sheet, usually glass, thick aluminum foil, or plastic. The resultant plate is dried and activated by heating in an oven for thirty minutes at 110 °C. The thickness of the absorbent layer is typically around 0.1–0.25 mm for analytical purposes and around 0.5–2.0 mm for preparative TLC. Other adsorbent coatings include aluminium oxide (alumina), or cellulose.

Applications

Reaction Monitoring and Characterization

TLC is a useful tool for reaction monitoring. For this, the plate normally contains a spot of starting material, a spot from the reaction mixture, and a co-spot (or cross-spot) containing both. The analysis will show if the starting material disappeared and if any new products appeared. This provides a quick and easy way to estimate how far a reaction has proceeded. In one study, TLC has been applied in the screening of organic reactions. The researchers react an alcohol and a catalyst directly in the co-spot of a TLC plate before developing it. This provides quick and easy small-scale testing of different reagents.

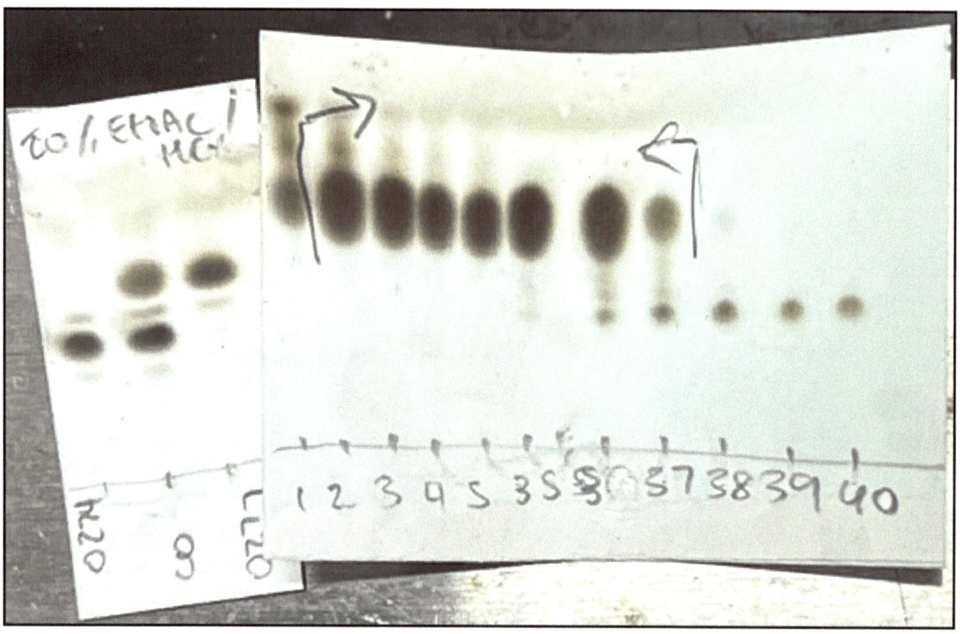

Fig: TLC for Reaction Monitoring and Choosing a Purification Solvent Mixture (Left) TLC from the Resulting Flash Column Chromatography (Right)

Compound characterization with TLC is also possible and is similar to reaction monitoring. However, rather than spotting with starting material and reaction mixture, it is with an unknown and a known compound. They may be the same compound if both spots have the same R_f and look the same under the chosen visualization method. However, co-elution complicates both reaction monitoring and characterization. This is because different compounds will move to the same spot on the plate. In such cases, different solvent mixtures may provide better separation.

Purity and Purification

Fig: Step 1

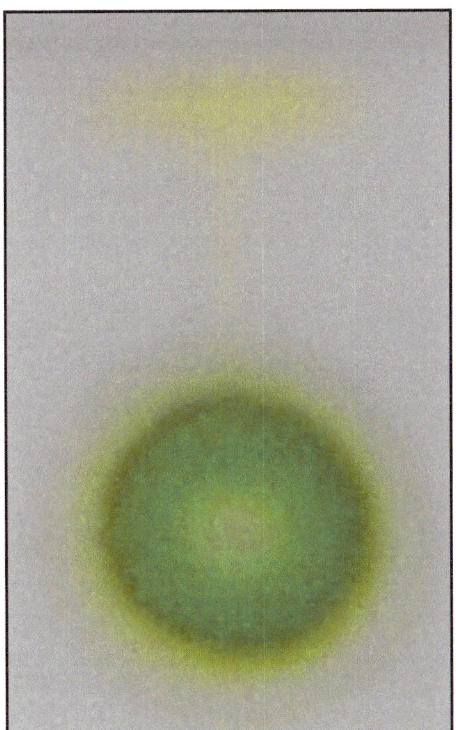

Fig: Step 2

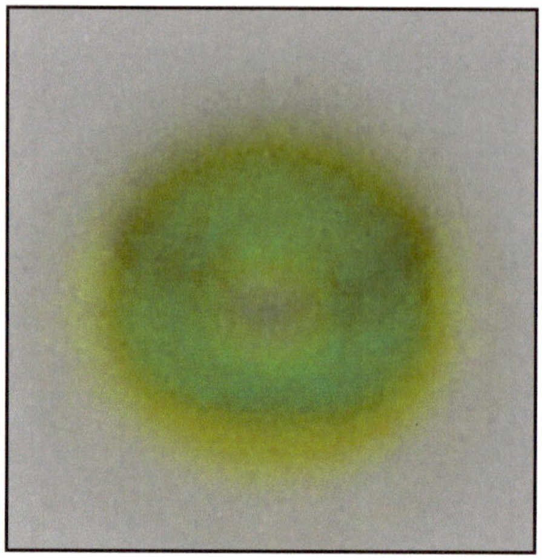

Fig: Step 3

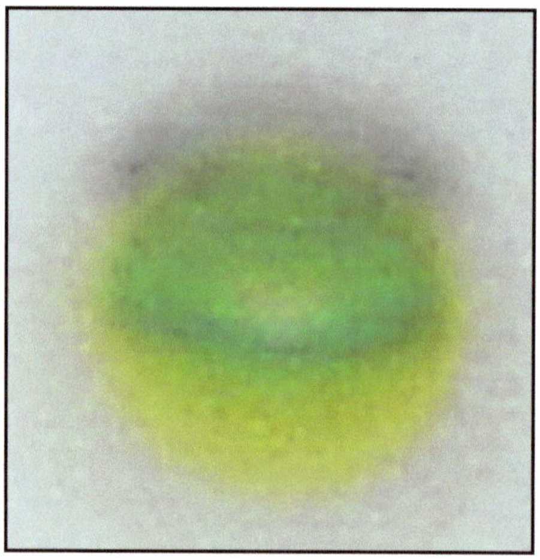

Fig: Step 4

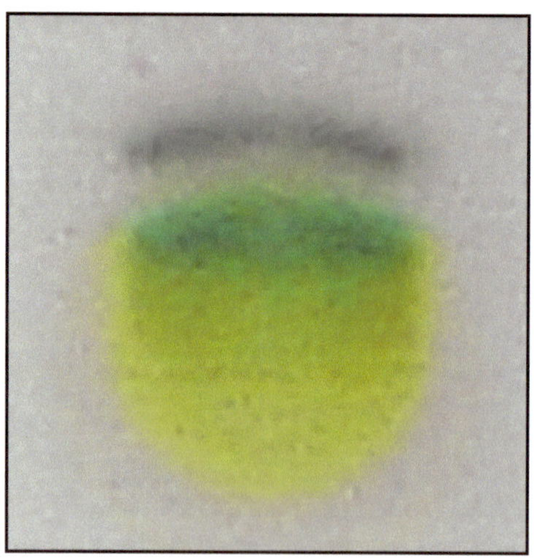

Fig: Step 5

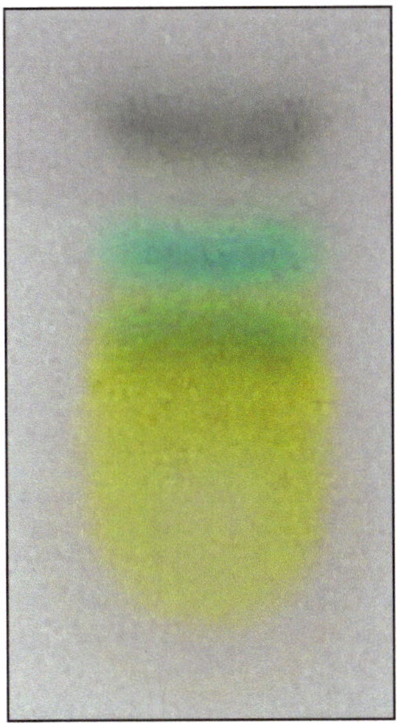

Fig: Step 6

Fig: Step 7

TLC helps show the purity of a sample. A pure sample should only contain one spot by TLC. TLC is also useful for small-scale purification. Because the separated compounds will be on different areas of the plate, a scientist can scrape off the stationary phase particles containing the desired compound and dissolve them into an appropriate solvent. Once all the compound dissolves in the solvent, they filter out the silica particles, then evaporate the solvent to isolate the product. Big preparative TLC plates with thick silica gel coatings can separate more than 100 mg of material.

For larger-scale purification and isolation, TLC is useful to quickly test solvent mixtures before running flash column chromatography on a large batch of impure material. A compound elutes from a column when the amount of solvent collected is equal to $1/R_f$. The eluent from flash column chromatography gets collected across several containers (for example, test tubes) called fractions. TLC helps show which fractions contain impurities and which contain pure compound.

Furthermore, two-dimensional TLC can help check if a compound is stable on a particular stationary phase. This test requires two runs on a square-shaped TLC plate. The plate is rotated by 90º before the second run. If the target compound appears on the diagonal of the square, it is stable on the chosen stationary phase. Otherwise, it is decomposing on the plate. If this is the case, an alternative stationary phase may prevent this decomposition. TLC is also an analytical method for the direct separation of enantiomers and the control of enantiomeric purity, e.g. active pharmaceutical ingredients (APIs) that are chiral.

Separation of green plant matter in spinach (note that images from steps 1-6 are zoomed into the bottom of the plate)

9.5 DISPLACEMENT CHROMATOGRAPHY

The basic principle of displacement chromatography is: A molecule with a high affinity for the chromatography matrix (the displacer) competes effectively for binding sites, and thus displaces all molecules with lesser affinities. There are distinct differences between displacement and elution chromatography. In elution mode, substances typically emerge from a column in narrow, Gaussian peaks. Wide separation of peaks, preferably to baseline, is desired for maximum purification. The speed at which any component of a mixture travels down the column in elution mode depends on many factors. But for two substances to travel at different speeds, and thereby be resolved, there must be substantial differences in some interaction between the biomolecules and the chromatography matrix. Operating parameters are adjusted to maximize the effect of this difference. In many cases, baseline separation of the peaks can be achieved only with gradient elution and low column loadings. Thus, two drawbacks to elution mode chromatography, especially at the preparative scale, are operational complexity, due to gradient solvent pumping, and low throughput, due to low column loadings. Displacement chromatography has advantages over elution chromatography in that components are resolved into consecutive zones of pure substances rather than "peaks". Because the process takes advantage of the nonlinearity of the isotherms, a larger column feed can be separated on a given column with the purified components recovered at significantly higher concentrations.

9.6 GAS CHROMATOGRAPHY

Gas chromatography (GC) is a common type of chromatography used in analytical chemistry for separating and analyzing compounds that can be vaporized without decomposition. Typical uses of GC include testing the purity of a particular substance, or separating the different components of a mixture. In preparative chromatography, GC can be used to prepare pure compounds from a mixture.

Gas chromatography is also sometimes known as vapor-phase chromatography (VPC), or gas-liquid partition chromatography (GLPC). These alternative names, as well as their respective abbreviations, are frequently used in scientific literature.

Gas chromatography is the process of separating compounds in a mixture by injecting a gaseous or liquid sample into a mobile phase, typically called the carrier gas, and passing the gas through a stationary phase. The mobile phase is usually an inert gas or an unreactive gas such as helium, argon, nitrogen or hydrogen. The stationary phase can be solid or liquid, although most GC systems today use a polymeric liquid stationary phase.

The stationary phase is contained inside of a separation column. Today, most GC columns are fused silica capillaries with an inner diameter of 100–320 micrometres (0.0039–0.0126 in) and a length of 5–60 metres (16–197 ft). The GC column is located inside an oven where the temperature of the gas can be controlled and the effluent coming off the column is monitored by a suitable detector.

9.6.1 Operating Principle

A gas chromatograph is made of a narrow tube, known as the column, through which the vaporized sample passes, carried along by a continuous flow of inert or nonreactive gas.

Components of the sample pass through the column at different rates, depending on their chemical and physical properties and the resulting interactions with the column lining or filling, called the stationary phase. The column is typically enclosed within a temperature controlled oven. As the chemicals exit the end of the column, they are detected and identified electronically.

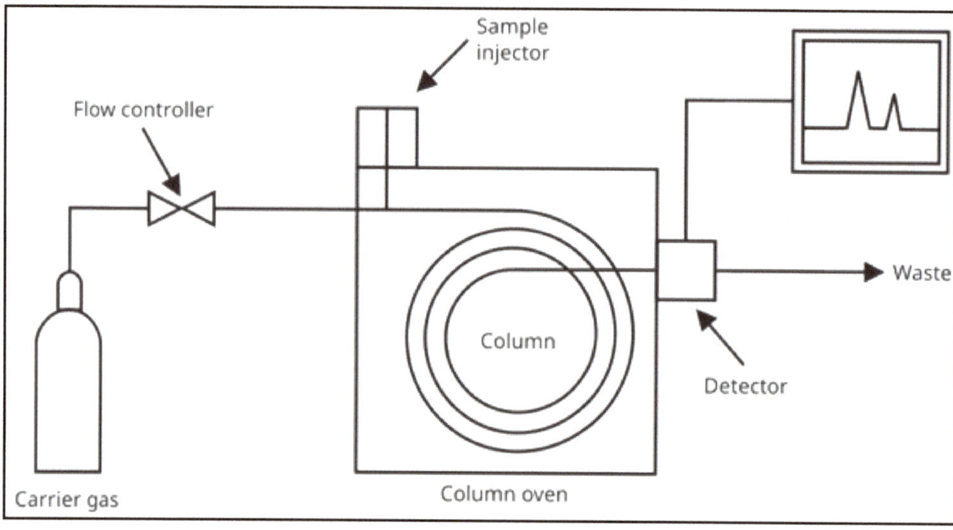

Fig: Diagram of a Gas Chromatograph

9.6.2 History

Fig: Gas Chromatograph

Background

Chromatography dates to 1903 in the work of the Russian scientist, Mikhail Semenovich Tswett, who separated plant pigments via liquid column chromatography.

Invention

The invention of gas chromatography is generally attributed to Anthony T. James and Archer J.P. Martin. Their gas chromatograph used partition chromatography as the separating principle, rather than adsorption chromatography. The popularity of gas chromatography quickly rose after the development of the flame ionization detector. Martin and another one of their colleagues, Richard Synge, with whom he shared the 1952 Nobel Prize in Chemistry, had noted in an earlier paper that chromatography might also be used to separate gases. Synge pursued other work while Martin continued his work with James.

Gas Adsorption Chromatography Precursors

German physical chemist Erika Cremer in 1947 together with Austrian graduate student Fritz Prior developed what could be considered the first gas chromatograph that consisted of a carrier gas, a column packed with silica gel, and a thermal conductivity detector. They exhibited the chromatograph at ACHEMA in Frankfurt, but nobody was interested in it. N.C. Turner with the Burrell Corporation introduced in 1943 a massive instrument that used a charcoal column and mercury vapors. Stig Claesson of Uppsala University published in 1946 his work on a charcoal column that also used mercury. Gerhard Hesse, while a professor at the University of Marburg/Lahn decided to test the prevailing opinion among German chemists that molecules could not be separated in a moving gas stream. He set up a simple glass column filled with starch and successfully separated bromine and iodine using nitrogen as the carrier gas. He then built a system that flowed an inert gas through a glass condenser packed with silica gel and collected the eluted fractions. Courtenay S.G Phillips of Oxford University investigated separation in a charcoal column using a thermal conductivity detector. He consulted with Claesson and decided to use displacement as his separating principle. After learning about the results of James and Martin, he switched to partition chromatography.

Column Technology

Early gas chromatography used packed columns, made of block 1–5 m long, 1–5 mm diameter, and filled with particles. The resolution of packed columns was improved by the invention of capillary column, in which the stationary phase is coated on the inner wall of the capillary.

9.6.3 Physical Components

Autosamplers

The autosampler provides the means to introduce a sample automatically into the inlets. Manual insertion of the sample is possible but is no longer common. Automatic insertion provides better reproducibility and time-optimization.

Different kinds of autosamplers exist. Autosamplers can be classified in relation to sample capacity (auto-injectors vs. autosamplers, where auto-injectors can work a small number of samples), to robotic technologies (XYZ robot vs. rotating robot – the most common), or to analysis:

- Liquid.
- Static head-space by syringe technology.
- Dynamic head-space by transfer-line technology.
- Solid phase microextraction (SPME).

Inlets

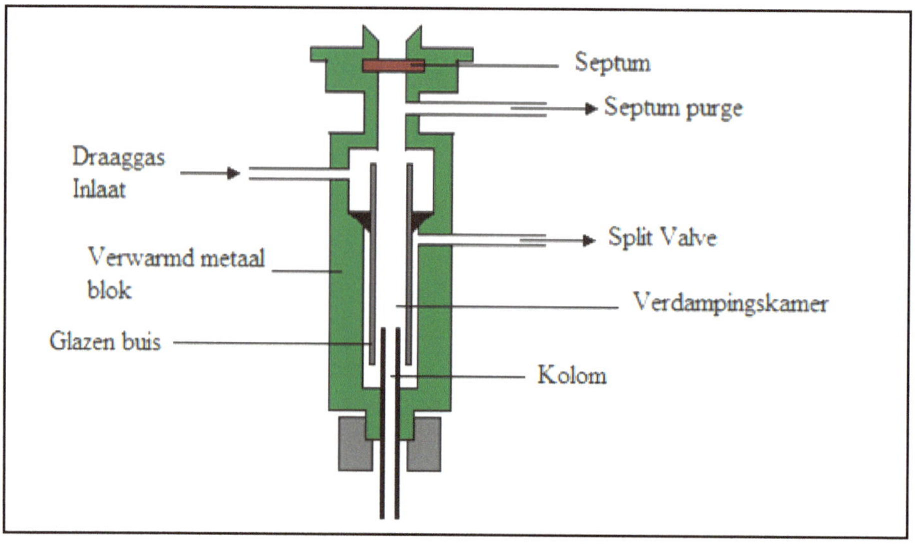

Fig: Split/Splitless Inlet

The column inlet (or injector) provides the means to introduce a sample into a continuous flow of carrier gas. The inlet is a piece of hardware attached to the column head.

Common inlet types are:

- **S/SL (Split/Splitless) Injector:** A sample is introduced into a heated small chamber via a syringe through a septum – the heat facilitates volatilization of the sample and sample matrix. The carrier gas then either sweeps the entirety (splitless mode) or a portion (split mode) of the sample into the column. In split mode, a part of the sample/carrier gas mixture in the injection chamber is exhausted through the split vent. Split injection is preferred when working with samples with high analyte concentrations (>0.1%) whereas splitless injection is best suited for trace analysis with low amounts of analytes (<0.01%). In splitless mode the split valve opens after a pre-set amount of time to purge heavier elements that would otherwise contaminate the system. This pre-set (splitless) time should be optimized, the shorter time (e.g., 0.2 min) ensures less tailing but loss in response, the longer time (2 min) increases tailing but also signal.

- **On-Column Inlet:** The sample is here introduced directly into the column in its entirety without heat, or at a temperature below the boiling point of the solvent. The low temperature condenses the sample into a narrow zone. The column and inlet can then be heated, releasing the sample into the gas phase. This ensures the lowest possible temperature for chromatography and keeps samples from decomposing above their boiling point.
- **PTV Injector:** Temperature-programmed sample introduction was first described by Vogt in 1979. Originally Vogt developed the technique as a method for the introduction of large sample volumes (up to 250 µL) in capillary GC. Vogt introduced the sample into the liner at a controlled injection rate. The temperature of the liner was chosen slightly below the boiling point of the solvent. The low-boiling solvent was continuously evaporated and vented through the split line. Based on this technique, Poy developed the programmed temperature vaporising injector; PTV. By introducing the sample at a low initial liner temperature many of the disadvantages of the classic hot injection techniques could be circumvented.
- **Gas Source Inlet or Gas Switching Valve:** Gaseous samples in collection bottles are connected to what is most commonly a six-port switching valve. The carrier gas flow is not interrupted while a sample can be expanded into a previously evacuated sample loop. Upon switching, the contents of the sample loop are inserted into the carrier gas stream.
- **P/T (Purge-and-Trap) System:** An inert gas is bubbled through an aqueous sample causing insoluble volatile chemicals to be purged from the matrix. The volatiles are 'trapped' on an absorbent column (known as a trap or concentrator) at ambient temperature. The trap is then heated and the volatiles are directed into the carrier gas stream. Samples requiring preconcentration or purification can be introduced via such a system, usually hooked up to the S/SL port.

The choice of carrier gas (mobile phase) is important. Hydrogen has a range of flow rates that are comparable to helium in efficiency. However, helium may be more efficient and provide the best separation if flow rates are optimized. Helium is non-flammable and works with a greater number of detectors and older instruments.

Therefore, helium is the most common carrier gas used. However, the price of helium has gone up considerably over recent years, causing an increasing number of chromatographers to switch to hydrogen gas. Historical use, rather than rational consideration, may contribute to the continued preferential use of helium.

Detectors

Commonly used detectors are the flame ionization detector (FID) and the thermal conductivity detector (TCD). While TCDs are beneficial in that they are non-destructive, its low detection limit for most analytes inhibits widespread use. FIDs are sensitive primarily to hydrocarbons, and are more sensitive to them than TCD. FIDs cannot detect water or carbon dioxide which make them ideal for environmental organic analyte analysis. FID is two to three times more sensitive to analyte detection than TCD.

The TCD relies on the thermal conductivity of matter passing around a thin wire of tungsten-rhenium with a current traveling through it. In this set up helium or nitrogen serve as the carrier gas because of their relatively high thermal conductivity which keep the filament cool and maintain uniform resistivity and electrical efficiency of the filament. When analyte molecules elute from the column, mixed with carrier gas, the thermal conductivity decreases while there is an increase in filament temperature and resistivity resulting in fluctuations in voltage ultimately causing a detector response. Detector sensitivity is proportional to filament current while it is inversely proportional to the immediate environmental temperature of that detector as well as flow rate of the carrier gas.

In a flame ionization detector (FID), electrodes are placed adjacent to a flame fueled by hydrogen / air near the exit of the column, and when carbon containing compounds exit the column they are pyrolyzed by the flame. This detector works only for organic / hydrocarbon containing compounds due to the ability of the carbons to form cations and electrons upon pyrolysis which generates a current between the electrodes. The increase in current is translated and appears as a peak in a chromatogram. FIDs have low detection limits (a few picograms per second) but they are unable to generate ions from carbonyl containing carbons. FID compatible carrier gasses include helium, hydrogen, nitrogen, and argon.

In FID, sometimes the stream is modified before entering the detector. A methanizer converts carbon monoxide and carbon dioxide into methane so that it can be detected. A different technology is the polyarc, by Activated Research Inc, that converts all compounds to methane.

Alkali flame detector (AFD) or alkali flame ionization detector (AFID) has high sensitivity to nitrogen and phosphorus, similar to NPD. However, the alkaline metal ions are supplied with the hydrogen gas, rather than a bead above the flame. For this reason AFD does not suffer the "fatigue" of the NPD, but provides a constant sensitivity over long period of time. In addition, when alkali ions are not added to the flame, AFD operates like a standard FID. A catalytic combustion detector (CCD) measures combustible hydrocarbons and hydrogen. Discharge ionization detector (DID) uses a high-voltage electric discharge to produce ions.

Flame photometric detector (FPD) uses a photomultiplier tube to detect spectral lines of the compounds as they are burned in a flame. Compounds eluting off the column are carried into a hydrogen fueled flame which excites specific elements in the molecules, and the excited elements (P,S, Halogens, Some Metals) emit light of specific characteristic wavelengths. The emitted light is filtered and detected by a photomultiplier tube. In particular, phosphorus emission is around 510–536 nm and sulfur emission is at 394 nm. With an atomic emission detector (AED), a sample eluting from a column enters a chamber which is energized by microwaves that induce a plasma. The plasma causes the analyte sample to decompose and certain elements generate an atomic emission spectra. The atomic emission spectra is diffracted by a diffraction grating and detected by a series of photomultiplier tubes or photo diodes.

Electron capture detector (ECD) uses a radioactive beta particle (electron) source to measure the degree of electron capture. ECD are used for the detection of molecules containing electronegative / withdrawing elements and functional groups like halogens, carbonyl, nitriles, nitro groups, and organometalics. In this type of detector either nitrogen or 5% methane in argon is used as the mobile phase carrier gas. The carrier gas passes between two electrodes placed at the end of the column, and adjacent to the cathode (negative electrode) resides a radioactive foil such as 63Ni. The radioactive foil emits a beta particle (electron) which collides with and ionizes the carrier gas to generate more ions resulting in a current. When analyte molecules with electronegative / withdrawing elements or functional groups electrons are captured which results in a decrease in current generating a detector response.

Nitrogen–phosphorus detector (NPD), a form of thermionic detector where nitrogen and phosphorus alter the work function on a specially coated bead and a resulting current is measured.

Dry electrolytic conductivity detector (DELCD) uses an air phase and high temperature (v. Coulsen) to measure chlorinated compounds.

Mass spectrometer (MS), also called GC-MS; highly effective and sensitive, even in a small quantity of sample. This detector can be used to identify the analytes in chromatograms by their mass spectrum. Some GC-MS are connected to an NMR spectrometer which acts as a backup detector. This combination is known as GC-MS-NMR. Some GC-MS-NMR are connected to an infrared spectrophotometer which acts as a backup detector. This combination is known as GC-MS-NMR-IR. It must, however, be stressed this is very rare as most analyses needed can be concluded via purely GC-MS.

Vacuum ultraviolet (VUV) represents the most recent development in gas chromatography detectors. Most chemical species absorb and have unique gas phase absorption cross sections in the approximately 120–240 nm VUV wavelength range monitored. Where absorption cross sections are known for analytes, the VUV detector is capable of absolute determination (without calibration) of the number of molecules present in the flow cell in the absence of chemical interferences.

Olfactometric detector, also called GC-O, uses a human assessor to analyse the odour activity of compounds. With an odour port or a sniffing port, the quality of the odour, the intensity of the odour and the duration of the odour activity of a compound can be assessed.

Other detectors include the Hall electrolytic conductivity detector (ElCD), helium ionization detector (HID), infrared detector (IRD), photo-ionization detector (PID), pulsed discharge ionization detector (PDD), and thermionic ionization detector (TID).

9.6.4 Methods

The method is the collection of conditions in which the GC operates for a given analysis. Method development is the process of determining what conditions are adequate and/or ideal for the analysis required.

Conditions which can be varied to accommodate a required analysis include inlet temperature, detector temperature, column temperature and temperature program, carrier gas and carrier gas flow rates, the column's stationary phase, diameter and length, inlet type and flow rates, sample size and injection technique. Depending on the detector(s) installed on the GC, there may be a number of detector conditions that can also be varied. Some GCs also include valves which can change the route of sample and carrier flow. The timing of the opening and closing of these valves can be important to method development.

Carrier Gas Selection and Flow Rates

Typical carrier gases include helium, nitrogen, argon, and hydrogen. Which gas to use is usually determined by the detector being used, for example, a DID requires helium as the carrier gas.

When analyzing gas samples the carrier is also selected based on the sample's matrix, for example, when analyzing a mixture in argon, an argon carrier is preferred because the argon in the sample does not show up on the chromatogram. Safety and availability can also influence carrier selection.

The purity of the carrier gas is also frequently determined by the detector, though the level of sensitivity needed can also play a significant role. Typically, purities of 99.995% or higher are used. The most common purity grades required by modern instruments for the majority of sensitivities are 5.0 grades, or 99.999% pure meaning that there is a total of 10 ppm of impurities in the carrier gas that could affect the results. The highest purity grades in common use are 6.0 grades, but the need for detection at very low levels in some forensic and environmental applications has driven the need for carrier gases at 7.0 grade purity and these are now commercially available. Trade names for typical purities include "Zero Grade", "Ultra-High Purity (UHP) Grade", "4.5 Grade" and "5.0 Grade".

The carrier gas linear velocity affects the analysis in the same way that temperature does. The higher the linear velocity the faster the analysis, but the lower the separation between analytes. Selecting the linear velocity is therefore the same compromise between the level of separation and length of analysis as selecting the column temperature. The linear velocity will be implemented by means of the carrier gas flow rate, with regards to the inner diameter of the column.

With GCs made before the 1990s, carrier flow rate was controlled indirectly by controlling the carrier inlet pressure, or "column head pressure". The actual flow rate was measured at the outlet of the column or the detector with an electronic flow meter, or a bubble flow meter, and could be an involved, time consuming, and frustrating process. It was not possible to vary the pressure setting during the run, and thus the flow was essentially constant during the analysis. The relation between flow rate and inlet pressure is calculated with Poiseuille's equation for compressible fluids.

Many modern GCs, however, electronically measure the flow rate, and electronically control the carrier gas pressure to set the flow rate. Consequently, carrier pressures and flow rates can be adjusted during the run, creating pressure/flow programs similar to temperature programs.

Stationary Compound Selection

The polarity of the solute is crucial for the choice of stationary compound, which in an optimal case would have a similar polarity as the solute. Common stationary phases in open tubular columns are cyanopropylphenyl dimethyl polysiloxane, carbowax polyethyleneglycol, biscyanopropyl cyanopropylphenyl polysiloxane and diphenyl dimethyl polysiloxane. For packed columns more options are available.

Inlet Types and Flow Rates

The choice of inlet type and injection technique depends on if the sample is in liquid, gas, adsorbed, or solid form, and on whether a solvent matrix is present that has to be vaporized. Dissolved samples can be introduced directly onto the column via a COC injector, if the conditions are well known; if a solvent matrix has to be vaporized and partially removed, a S/SL injector is used (most common injection technique); gaseous samples (e.g., air cylinders) are usually injected using a gas switching valve system; adsorbed samples (e.g., on adsorbent tubes) are introduced using either an external (on-line or off-line) desorption apparatus such as a purge-and-trap system, or are desorbed in the injector (SPME applications).

Sample Size and Injection Technique

Sample Injection

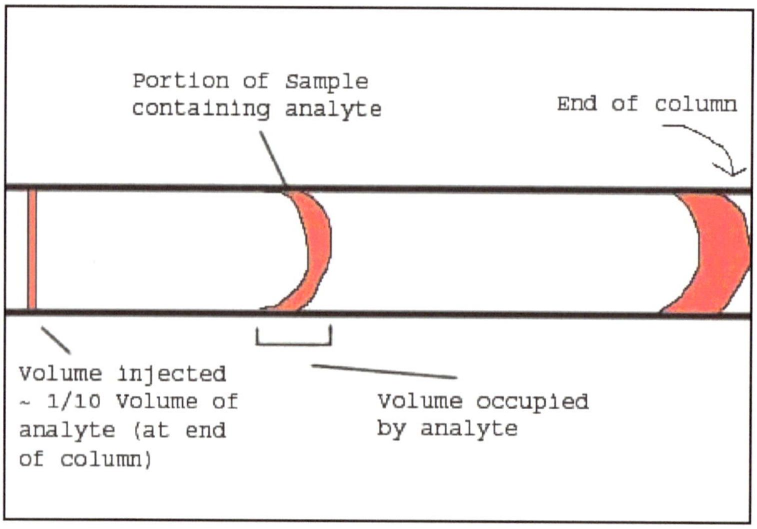

Fig: The Rule of Ten in Gas Chromatography

The real chromatographic analysis starts with the introduction of the sample onto the column. The development of capillary gas chromatography resulted in many practical problems with the injection technique. The technique of on-column injection, often used with packed columns, is usually not possible with capillary columns. In the injection system in the capillary gas chromatograph the amount injected should not overload the column and the width of the injected plug should be small compared to the spreading due to the chromatographic process. Failure to comply with this latter requirement will reduce the separation capability of the column. As a general rule, the volume injected, V_{inj}, and the volume of the detector cell, V_{det}, should be about 1/10 of the volume occupied by the portion of sample containing the molecules of interest (analytes) when they exit the column.

Some general requirements which a good injection technique should fulfill are that it should be possible to obtain the column's optimum separation efficiency, it should allow accurate and reproducible injections of small amounts of representative samples, it should induce no change in sample composition, it should not exhibit discrimination based on differences in boiling point, polarity, concentration or thermal/catalytic stability, and it should be applicable for trace analysis as well as for undiluted samples.

However, there are a number of problems inherent in the use of syringes for injection. Even the best syringes claim an accuracy of only 3%, and in unskilled hands, errors are much larger.

The needle may cut small pieces of rubber from the septum as it injects sample through it. These can block the needle and prevent the syringe filling the next time it is used. It may not be obvious that this has happened. A fraction of the sample may get trapped in the rubber, to be released during subsequent injections. This can give rise to ghost peaks in the chromatogram. There may be selective loss of the more volatile components of the sample by evaporation from the tip of the needle.

Column Selection

The choice of column depends on the sample and the active measured. The main chemical attribute regarded when choosing a column is the polarity of the mixture, but functional groups can play a large part in column selection. The polarity of the sample must closely match the polarity of the column stationary phase to increase resolution and separation while reducing run time.

The separation and run time also depends on the film thickness (of the stationary phase), the column diameter and the column length.

Column Temperature and Temperature Program

The column(s) in a GC are contained in an oven, the temperature of which is precisely controlled electronically. The rate at which a sample passes through the column is directly proportional to the temperature of the column.

The higher the column temperature, the faster the sample moves through the column. However, the faster a sample moves through the column, the less it interacts with the stationary phase, and the less the analytes are separated.

In general, the column temperature is selected to compromise between the length of the analysis and the level of separation.

A method which holds the column at the same temperature for the entire analysis is called "isothermal". Most methods, however, increase the column temperature during the analysis, the initial temperature, rate of temperature increase (the temperature "ramp"), and final temperature are called the temperature program.

A temperature program allows analytes that elute early in the analysis to separate adequately, while shortening the time it takes for late-eluting analytes to pass through the column.

9.6.5 Data Reduction and Analysis

Qualitative Analysis

Generally, chromatographic data is presented as a graph of detector response (y-axis) against retention time (x-axis), which is called a chromatogram. This provides a spectrum of peaks for a sample representing the analytes present in a sample eluting from the column at different times. Retention time can be used to identify analytes if the method conditions are constant. Also, the pattern of peaks will be constant for a sample under constant conditions and can identify complex mixtures of analytes. However, in most modern applications, the GC is connected to a mass spectrometer or similar detector that is capable of identifying the analytes represented by the peaks.

Quantitative Analysis

The area under a peak is proportional to the amount of analyte present in the chromatogram. By calculating the area of the peak using the mathematical function of integration, the concentration of an analyte in the original sample can be determined. Concentration can be calculated using a calibration curve created by finding the response for a series of concentrations of analyte, or by determining the relative response factor of an analyte. The relative response factor is the expected ratio of an analyte to an internal standard (or external standard) and is calculated by finding the response of a known amount of analyte and a constant amount of internal standard (a chemical added to the sample at a constant concentration, with a distinct retention time to the analyte). In most modern GC-MS systems, computer software is used to draw and integrate peaks, and match MS spectra to library spectra.

9.6.6 Applications

In general, substances that vaporize below 300 °C (and therefore are stable up to that temperature) can be measured quantitatively. The samples are also required to be salt-free; they should not contain ions. Very minute amounts of a substance can be measured, but it is often required that the sample must be measured in comparison to a sample containing the pure, suspected substance known as a reference standard.

Various temperature programs can be used to make the readings more meaningful; for example to differentiate between substances that behave similarly during the GC process. Professionals working with GC analyze the content of a chemical product, for example in assuring the quality of products in the chemical industry; or measuring chemicals in soil, air or water, such as soil gases. GC is very accurate if used properly and can measure picomoles of a substance in a 1 ml liquid sample, or parts-per-billion concentrations in gaseous samples.

In practical courses at colleges, students sometimes get acquainted to the GC by studying the contents of lavender oil or measuring the ethylene that is secreted by Nicotiana benthamiana plants after artificially injuring their leaves. These GC analyse hydrocarbons (C2-C40+). In a typical experiment, a packed column is used to separate the light gases, which are then detected with a TCD. The hydrocarbons are separated using a capillary column and detected with a FID. A complication with light gas analyses that include H_2 is that He, which is the most common and most sensitive inert carrier (sensitivity is proportional to molecular mass) has an almost identical thermal conductivity to hydrogen (it is the difference in thermal conductivity between two separate filaments in a Wheatstone Bridge type arrangement that shows when a component has been eluted). For this reason, dual TCD instruments used with a separate channel for hydrogen that uses nitrogen as a carrier are common. Argon is often used when analysing gas phase chemistry reactions such as F-T synthesis so that a single carrier gas can be used rather than two separate ones. The sensitivity is reduced, but this is a trade off for simplicity in the gas supply.

Gas chromatography is used extensively in forensic science. Disciplines as diverse as solid drug dose (pre-consumption form) identification and quantification, arson investigation, paint chip analysis, and toxicology cases, employ GC to identify and quantify various biological specimens and crime-scene evidence.

9.7 LIQUID CHROMATOGRAPHY

Liquid chromatography (LC) is a separation technique in which the mobile phase is a liquid. It can be carried out either in a column or a plane. Present day liquid chromatography that generally utilizes very small packing particles and a relatively high pressure is referred to as high-performance liquid chromatography.

In HPLC the sample is forced by a liquid at high pressure (the mobile phase) through a column that is packed with a stationary phase composed of irregularly or spherically shaped particles, a porous monolithic layer, or a porous membrane. Monoliths are "sponge-like chromatographic media" and are made up of an unending block of organic or inorganic parts. HPLC is historically divided into two different sub-classes based on the polarity of the mobile and stationary phases. Methods in which the stationary phase is more polar than the mobile phase (e.g., toluene as the mobile phase, silica as the stationary phase) are termed normal phase liquid chromatography (NPLC) and the opposite (e.g., water-methanol mixture as the mobile phase and C18 (octadecylsilyl) as the stationary phase) is termed reversed phase liquid chromatography (RPLC).

9.8 SUPERCRITICAL FLUID CHROMATOGRAPHY

Supercritical fluid chromatography (SFC) is a form of normal phase chromatography that uses a supercritical fluid such as carbon dioxide as the mobile phase. It is used for the analysis and purification of low to moderate molecular weight, thermally labile molecules and can also be used for the separation of chiral compounds. Principles are similar to those of high performance liquid chromatography (HPLC); however, SFC typically utilizes carbon dioxide as the mobile phase.

Therefore, the entire chromatographic flow path must be pressurized. Because the supercritical phase represents a state whereby bulk liquid and gas properties converge, supercritical fluid chromatography is sometimes called convergence chromatography. The idea of liquid and gas properties convergence was first envisioned by Giddings.

9.8.1 Applications

SFC has been used primarily for separation of chiral molecules, mainly those which required normal phase conditions. While the mobile phase is a fluid in the supercritical state, the stationary phase is packed inside columns similar to those used in liquid chromatography. Since the use of normal phase mode of chromatography remained less common, so did SFC; therefore it is now commonly used for selected chiral and achiral separations and purification in the pharmaceutical industry.

9.8.2 Apparatus

Instrumentation of supercritical fluid chromatography SFC has a similar setup to an HPLC instrument. The stationary phases are similar, and are packed inside similar column types. However, there are special features in these systems, because of the need to keep the mobile phase at supercritical fluidic state over the entire system.

Temperature is critical to keep the fluids in a supercritical state, so there should be a heat control tool in the system, similar to that of GC. Also, there should be a precise pressure control mechanism, a restrictor to keep the pressure above a certain point, because pressure is another essential parameter to keep the mobile phase in a supercritical fluid state, so it is kept at the required minimal level. A microprocessor mechanism is placed in the instrument for SFC. This unit collects data for pressure, oven temperature, and detector performance to control the related pieces of the instrument.

CO_2 utilized in carbon dioxide dedicated pumps, which require that the incoming CO_2 and pump heads be kept cold, in order to maintain the carbon dioxide at a temperature and pressure fit for supercritical fluidic state, where it can be effectively metered at a specified flow rate range.

The CO_2 subsequently becomes supercritical fluid throughout the injector and the column oven, when the temperature and pressure it is subjected to, are raised above the critical point of the liquid, thus the supercritical state is achieved. Supercritical fluids combine useful properties of gas and liquid phases, as it can behave like both a gas and a liquid in various aspects. A supercritical fluid provides a gas-like characteristic when it fills a container and it takes the shape of the container. The motion and kinetics of the molecules are quite similar to gas molecules. On the other hand, a supercritical fluid behaves like a liquid because its density property is near liquid; thus, a supercritical fluid shows a similarity to the dissolving effect of a liquid.

The result is that one can load masses, similar to those used in HPLC, on column per injection, and still maintain a high chromatographic efficiency similar to those attained in GC. Typically, gradient elution is employed in analytical SFC using a polar co-solvent such as methanol, possibly with a weak acid or base at low concentrations ~1%. The apparent plate count per analysis can be observed to exceed 500K plates per meter routinely with 5 um stationary phases. The operator uses software to set mobile phase flow rate, co-solvent composition, system back pressure and column oven temperature, which must exceed 40 °C for supercritical conditions needed to be achieved with CO_2. In addition, SFC provides an additional control parameter - pressure - by using an automated static and dynamic back pressure regulator. From an operational standpoint, SFC is as simple and robust as HPLC, but fraction collection is more convenient because the primary mobile phase evaporates leaving only the analyte and a small volume of polar co-solvent. If the outlet CO_2 is captured, it can be re-compressed and recycled, allowing for >90% reuse of CO_2.

Similar to HPLC, SFC uses a variety of detection methods including UV/VIS, mass spectrometry, FID (unlike HPLC) and evaporative light scattering.

9.8.3 Sample Preparation

A rule-of-thumb is that any molecule that will dissolve in methanol or a less polar solvent is compatible with SFC, including non-volatile polar solutes. CO_2 has polarity similar to n-heptane at its critical point. The solvent's elution strength can be increased just by increasing density or alternatively, using a polar co-solvent. In practice, when the fraction of the co-solvent is high, the mobile phase might not be truly at supercritical fluid state, but this terminology is used regardless, and the chromatograms show better elution and higher efficiency nevertheless.

9.8.4 Mobile Phase

The mobile phase is composed primarily of supercritical carbon dioxide, but since CO_2 on its own is too non-polar to effectively elute many analytes, cosolvents are added to modify the mobile phase polarity. Cosolvents are typically simple alcohols like methanol, ethanol, or isopropyl alcohol. Other solvents such as acetonitrile, chloroform, or ethyl acetate can be used as modifiers. For food-grade materials, the selected cosolvent is often ethanol or ethyl acetate, both of which are generally recognized as safe (GRAS). The solvent limitations are system and column based.

9.8.5 Drawbacks

There have been a few technical issues that have limited adoption of SFC technology in the past. First of all, is the need to keep a high gas pressure in the operating conditions. High-pressure vessels are expensive and bulky, and special materials are often needed to avoid dissolving gaskets and O-rings in the supercritical fluid. A second drawback is difficulty in maintaining pressure constant (by back-pressure regulation). Whereas liquids are nearly incompressible, so their densities are constant regardless of pressure, supercritical fluids are highly compressible and their physical properties change with pressure – such as the pressure drop across a packed-bed column. Currently, automated backpressure regulators can maintain a constant pressure in the column even if flow rate varies, mitigating this problem. A third drawback is difficulty in gas/liquid separation during collection of product. Upon depressurization, the CO_2 rapidly turns into gas and aerosolizes any dissolved analyte in the process. Cyclone separators have lessened difficulties in gas/liquid separations.

9.9 AFFINITY CHROMATOGRAPHY

Affinity chromatography is a method of separating a biomolecule from a mixture, based on a highly specific macromolecular binding interaction between the biomolecule and another substance. The specific type of binding interaction depends on the biomolecule of interest; antigen and antibody, enzyme and substrate, receptor and ligand, or protein and nucleic acid binding interactions are frequently exploited for isolation of various biomolecules. Affinity chromatography is useful for its high selectivity and resolution of separation, compared to other chromatographic methods.

9.9.1 Principle

Affinity chromatography has the advantage of specific binding interactions between the analyte of interest (normally dissolved in the mobile phase), and a binding partner or ligand (immobilized on the stationary phase). In a typical affinity chromatography experiment, the ligand is attached to a solid, insoluble matrix—usually a polymer such as agarose or polyacrylamide—chemically modified to introduce reactive functional groups with which the ligand can react, forming stable covalent bonds. The stationary phase is first loaded into a column to which the mobile phase is introduced. Molecules that bind to the ligand will remain associated with the stationary phase. A wash buffer is then applied to remove non-target biomolecules by disrupting their weaker interactions with the stationary phase, while the biomolecules of interest will remain bound. Target biomolecules may then be removed by applying a so-called elution buffer, which disrupts interactions between the bound target biomolecules and the ligand. The target molecule is thus recovered in the eluting solution.

Affinity chromatography does not require the molecular weight, charge, hydrophobicity, or other physical properties of the analyte of interest to be known, although knowledge of its binding properties is useful in the design of a separation protocol. Types of binding interactions commonly exploited in affinity chromatography procedures are summarized in the table.

Table: Typical Biological Interactions Used in Affinity Chromatography

Sr. No	Types of Ligand	Target Molecule
1	Substrate analogue	Enzymes
2	Antibody	Antigen
3	Lectin	Polysaccharide
4	Nucleic acid	Complementary base sequence
5	Hormone	Receptor
6	Avidin	Biotin/Biotin-conjugated molecule
7	Calmodulin	Calmodulin binding partner
8	Glutathione	GST fusion protein
9	Protein A or Protein G	Immunoglobulins
10	Nickel-NTA	polyhistidine fusion protein

9.9.2 Batch and Column Setups

Binding to the solid phase may be achieved by column chromatography whereby the solid medium is packed onto a column, the initial mixture run through the column to allow settling, a wash buffer run through the column and the elution buffer subsequently applied to the column and collected. These steps are usually done at ambient pressure.

Alternatively, binding may be achieved using a batch treatment, for example, by adding the initial mixture to the solid phase in a vessel, mixing, separating the solid phase, removing the liquid phase, washing, re-centrifuging, adding the elution buffer, re-centrifuging and removing the elute.

Sometimes a hybrid method is employed such that the binding is done by the batch method, but the solid phase with the target molecule bound is packed onto a column and washing and elution are done on the column.

The ligands used in affinity chromatography are obtained from both organic and inorganic sources. Examples of biological sources are serum proteins, lectins and antibodies. Inorganic sources are moronic acid, metal chelates and triazine dyes.

A third method, expanded bed absorption has also been developed. The solid phase particles are placed in a column where liquid phase is pumped in from the bottom and exits at the top. The gravity of the particles ensure that the solid phase does not exit the column with the liquid phase.

Affinity columns can be eluted by changing salt concentrations, pH, pI, charge and ionic strength directly or through a gradient to resolve the particles of interest.

More recently, setups employing more than one column in series have been developed. The advantage compared to single column setups is that the resin material can be fully loaded since non-binding product is directly passed on to a consecutive column with fresh column material. These chromatographic processes are known as periodic counter-current chromatography (PCC). The resin costs per amount of produced product can thus be drastically reduced. Since one column can always be eluted and regenerated while the other column is loaded, already two columns are sufficient to make full use of the advantages. Additional columns can give additional flexibility for elution and regeneration times, at the cost of additional equipment and resin costs.

9.9.3 Specific Uses

Affinity chromatography can be used in a number of applications, including nucleic acid purification, protein purification from cell free extracts, and purification from blood. By using affinity chromatography, one can separate proteins that bind to a certain fragment from proteins that do not bind that specific fragment. Because this technique of purification relies on the biological properties of the protein needed, it is a useful technique and proteins can be purified many folds in one step.

Various Affinity Media

Many different affinity media exist for a variety of possible uses. Briefly, they are (generalized) activated/functionalized that work as a functional spacer, support matrix, and eliminates handling of toxic reagents.

Amino acid media is used with a variety of serum proteins, proteins, peptides, and enzymes, as well as rRNA and dsDNA. Avidin biotin media is used in the purification process of biotin/avidin and their derivatives.

Carbohydrate bonding is most often used with glycoproteins or any other carbohydrate-containing substance; carbohydrate is used with lectins, glycoproteins, or any other carbohydrate metabolite protein. Dye ligand media is nonspecific but mimics biological substrates and proteins. Glutathione is useful for separation of GST tagged recombinant proteins. Heparin is a generalized affinity ligand, and it is most useful for separation of plasma coagulation proteins, along with nucleic acid enzymes and lipases

Nucleic acids function to trap mRNA, DNA, rRNA, and other nucleic acids/ oligonucleotides. Protein A/G method is used to purify immunoglobulins. Speciality media are designed for a specific class or type of protein/co enzyme; this type of media will only work to separate a specific protein or coenzyme.

Immunoaffinity

Another use for the procedure is the affinity purification of antibodies from blood serum. If the serum is known to contain antibodies against a specific antigen (for example if the serum comes from an organism immunized against the antigen concerned) then it can be used for the affinity purification of that antigen. This is also known as Immunoaffinity Chromatography. For example, if an organism is immunised against a GST-fusion protein it will produce antibodies against the fusion-protein, and possibly antibodies against the GST tag as well. The protein can then be covalently coupled to a solid support such as agarose and used as an affinity ligand in purifications of antibody from immune serum.

For thoroughness, the GST protein and the GST-fusion protein can each be coupled separately. The serum is initially allowed to bind to the GST affinity matrix. This will remove antibodies against the GST part of the fusion protein. The serum is then separated from the solid support and allowed to bind to the GST-fusion protein matrix. This allows any antibodies that recognize the antigen to be captured on the solid support. Elution of the antibodies of interest is most often achieved using a low pH buffer such as glycine pH 2.8. The eluate is collected into a neutral tris or phosphate buffer, to neutralize the low pH elution buffer and halt any degradation of the antibody's activity. This is a nice example as affinity purification is used to purify the initial GST-fusion protein, to remove the undesirable anti-GST antibodies from the serum and to purify the target antibody.

Monoclonal antibodies can also be selected to bind proteins with great specificity, where protein is released under fairly gentle conditions. This can become of use for further research in the future.

A simplified strategy is often employed to purify antibodies generated against peptide antigens. When the peptide antigens are produced synthetically, a terminal cysteine residue is added at either the N- or C-terminus of the peptide. This cysteine residue contains a sulfhydryl functional group which allows the peptide to be easily conjugated to a carrier protein (e.g. Keyhole limpet hemocyanin (KLH)). The same cysteine-containing peptide is also immobilized onto an agarose resin through the cysteine residue and is then used to purify the antibody.

Most monoclonal antibodies have been purified using affinity chromatography based on immunoglobulin-specific Protein A or Protein G, derived from bacteria. Immunoaffinity chromatography with monoclonal antibodies immobilized on monolithic column has been successfully used to capture extracellular vesicles (e.g., exosomes and exomeres) from human blood plasma by targeting tetraspanins and integrins found on the surface of the EVs. Immunoaffinity chromatography is also the basis for immunochromatographic test (ICT) strips, which provide a rapid means of diagnosis in patient care. Using ICT, a technician can make a determination at a patient's bedside, without the need for a laboratory. ICT detection is highly specific to the microbe causing an infection.

Immobilized Metal Ion Affinity Chromatography

Immobilized metal ion affinity chromatography (IMAC) is based on the specific coordinate covalent bond of amino acids, particularly histidine, to metals. This technique works by allowing proteins with an affinity for metal ions to be retained in a column containing immobilized metal ions, such as cobalt, nickel, or copper for the purification of histidine-containing proteins or peptides, iron, zinc or gallium for the purification of phosphorylated proteins or peptides. Many naturally occurring proteins do not have an affinity for metal ions, therefore recombinant DNA technology can be used to introduce such a protein tag into the relevant gene. Methods used to elute the protein of interest include changing the pH, or adding a competitive molecule, such as imidazole.

Recombinant Proteins

Possibly the most common use of affinity chromatography is for the purification of recombinant proteins. Proteins with a known affinity are protein tagged in order to aid their purification. The protein may have been genetically modified so as to allow it to be selected for affinity binding; this is known as a fusion protein. Protein tags include hexahistidine (His), glutathione-S-transferase (GST), maltose binding protein (MBP), and the Colicin E7 variant CL7 tag. Histidine tags have an affinity for nickel, cobalt, zinc, copper and iron ions which have been immobilized by forming coordinate covalent bonds with a chelator incorporated in the stationary phase. For elution, an excess amount of a compound able to act as a metal ion ligand, such as imidazole, is used. GST has an affinity for glutathione which is commercially available immobilized as glutathione agarose. During elution, excess glutathione is used to displace the tagged protein. CL7 has an affinity and specificity for Immunity Protein 7 (Im7) which is commercially available immobilized as Im7 agarose resin. For elution, an active and site-specific protease is applied to the Im7 resin to release the tag-free protein.

Lectins

Lectin affinity chromatography is a form of affinity chromatography where lectins are used to separate components within the sample. Lectins, such as concanavalin A are proteins which can bind specific alpha-D-mannose and alpha-D-glucose carbohydrate molecules. Some common carbohydrate molecules that is used in lectin affinity chromatography are Con A-Sepharose and WGA-agarose. Another example of a lectin is wheat germ agglutinin which binds D-N-acetyl-glucosamine. The most common application is to separate glycoproteins from non-glycosylated proteins, or one glycoform from another glycoform. Although there are various ways to perform lectin affinity chromatography, the goal is extract a sugar ligand of the desired protein.

Specialty

Another use for affinity chromatography is the purification of specific proteins using a gel matrix that is unique to a specific protein. For example, the purification of E. coli β-galactosidase is accomplished by affinity chromatography using p-aminobenyl-1-thio-β-D-galactopyranosyl agarose as the affinity matrix. p-aminobenyl-1-thio-β-D-galactopyranosyl agarose is used as the affinity matrix because it contains a galactopyranosyl group, which serves as a good substrate analog for E. coli β-Galactosidase. This property allows the enzyme to bind to the stationary phase of the affinity matrix and β-Galactosidase is eluted by adding increasing concentrations of salt to the column.

Alkaline Phosphatase

Alkaline phosphatase from E. coli can be purified using a DEAE-Cellulose matrix. A. phosphatase has a slight negative charge, allowing it to weakly bind to the positively charged amine groups in the matrix. The enzyme can then be eluted out by adding buffer with higher salt concentrations.

Boronate Affinity Chromatography

Boronate affinity chromatography consists of using boronic acid or boronates to elute and quantify amounts of glycoproteins. Clinical adaptations have applied this type of chromatography for use in determining long term assessment of diabetic patients through analysis of their glycated hemoglobin.

Serum Albumin Purification

Affinity purification of albumin and macroglobulin contamination is helpful in removing excess albumin and α_2-macroglobulin contamination, when performing mass spectrometry. In affinity purification of serum albumin, the stationary used for collecting or attracting serum proteins can be Cibacron Blue-Sepharose. Then the serum proteins can be eluted from the adsorbent with a buffer containing thiocyanate (SCN^-).

9.9.4 Weak Affinity Chromatography

Weak affinity chromatography (WAC) is an affinity chromatography technique for affinity screening in drug development. WAC is an affinity-based liquid chromatographic technique that separates chemical compounds based on their different weak affinities to an immobilized target. The higher affinity a compound has towards the target, the longer it remains in the separation unit, and this will be expressed as a longer retention time. The affinity measure and ranking of affinity can be achieved by processing the obtained retention times of analyzed compounds. Affinity chromatography is part of a larger suite of techniques used in chemoproteomics based drug target identification. The WAC technology is demonstrated against a number of different protein targets – proteases, kinases, chaperones and protein–protein interaction (PPI) targets. WAC has been shown to be more effective than established methods for fragment based screening.

9.10 SPECIAL TECHNIQUES

9.10.1 Reversed-Phase Chromatography

Reversed-phase chromatography (RPC) is any liquid chromatography procedure in which the mobile phase is significantly more polar than the stationary phase. It is so named because in normal-phase liquid chromatography, the mobile phase is significantly less polar than the stationary phase. Hydrophobic molecules in the mobile phase tend to adsorb to the relatively hydrophobic stationary phase. Hydrophilic molecules in the mobile phase will tend to elute first. Separating columns typically comprise a C8 or C18 carbon-chain bonded to a silica particle substrate.

9.10.2 Hydrophobic Interaction Chromatography

Hydrophobic Interaction Chromatography (HIC) is a purification and analytical technique that separates analytes, such as proteins, based on hydrophobic interactions between that analyte and the chromatographic matrix. It can provide a non-denaturing orthogonal approach to reversed phase separation, preserving native structures and potentially protein activity. In hydrophobic interaction chromatography, the matrix material is lightly substituted with hydrophobic groups.

These groups can range from methyl, ethyl, propyl, butyl, octyl, or phenyl groups. At high salt concentrations, non-polar sidechains on the surface on proteins "interact" with the hydrophobic groups; that is, both types of groups are excluded by the polar solvent (hydrophobic effects are augmented by increased ionic strength).

Thus, the sample is applied to the column in a buffer which is highly polar, which drives an association of hydrophobic patches on the analyte with the stationary phase. The eluent is typically an aqueous buffer with decreasing salt concentrations, increasing concentrations of detergent (which disrupts hydrophobic interactions), or changes in pH. Of critical importance is the type of salt used, with more kosmotropic salts as defined by the Hofmeister series providing the most water structuring around the molecule and resulting hydrophobic pressure. Ammonium sulfate is frequently used for this purpose. The addition of organic solvents or other less polar constituents may assist in improving resolution.

In general, Hydrophobic Interaction Chromatography (HIC) is advantageous if the sample is sensitive to pH change or harsh solvents typically used in other types of chromatography but not high salt concentrations. Commonly, it is the amount of salt in the buffer which is varied. In 2012, Müller and Franzreb described the effects of temperature on HIC using Bovine Serum Albumin (BSA) with four different types of hydrophobic resin. The study altered temperature as to effect the binding affinity of BSA onto the matrix. It was concluded that cycling temperature from 40 to 10 degrees Celsius would not be adequate to effectively wash all BSA from the matrix but could be very effective if the column would only be used a few times. Using temperature to effect change allows labs to cut costs on buying salt and saves money.

If high salt concentrations along with temperature fluctuations want to be avoided one can use a more hydrophobic to compete with one's sample to elute it. This so-called salt independent method of HIC showed a direct isolation of Human Immunoglobulin G (IgG) from serum with satisfactory yield and used β-cyclodextrin as a competitor to displace IgG from the matrix. This largely opens up the possibility of using HIC with samples which are salt sensitive as we know high salt concentrations precipitate proteins.

9.10.3 Hydrodynamic Chromatography

Hydrodynamic chromatography (HDC) is derived from the observed phenomenon that large droplets move faster than small ones. In a column, this happens because the center of mass of larger droplets is prevented from being as close to the sides of the column as smaller droplets because of their larger overall size. Larger droplets will elute first from the middle of the column while smaller droplets stick to the sides of the column and elute last. This form of chromatography is useful for separating analytes by molar mass (or molecular mass), size, shape, and structure when used in conjunction with light scattering detectors, viscometers, and refractometers. The two main types of HDC are open tube and packed column. Open tube offers rapid separation times for small particles, whereas packed column HDC can increase resolution and is better suited for particles with an average molecular mass larger than 10 5 10^5 daltons. HDC differs from other types of chromatography because the separation only takes place in the interstitial volume, which is the volume surrounding and in between particles in a packed column.

HDC shares the same order of elution as Size Exclusion Chromatography (SEC) but the two processes still vary in many ways. In a study comparing the two types of separation, Isenberg, Brewer, Côté, and Striegel use both methods for polysaccharide characterization and conclude that HDC coupled with multiangle light scattering (MALS) achieves more accurate molar mass distribution when compared to off-line MALS than SEC in significantly less time. This is largely due to SEC being a more destructive technique because of the pores in the column degrading the analyte during separation, which tends to impact the mass distribution. However, the main disadvantage of HDC is low resolution of analyte peaks, which makes SEC a more viable option when used with chemicals that are not easily degradable and where rapid elution is not important.

HDC plays an especially important role in the field of microfluidics. The first successful apparatus for HDC-on-a-chip system was proposed by Chmela, et al. in 2002. Their design was able to achieve separations using an 80 mm long channel on the timescale of 3 minutes for particles with diameters ranging from 26 to 110 nm, but the authors expressed a need to improve the retention and dispersion parameters. In a 2010 publication by Jellema, Markesteijn, Westerweel, and Verpoorte, implementing HDC with a recirculating bidirectional flow resulted in high resolution, size based separation with only a 3 mm long channel. Having such a short channel and high resolution was viewed as especially impressive considering that previous studies used channels that were 80 mm in length. For a biological application, in 2007, Huh, et al. proposed a microfluidic sorting device based on HDC and gravity, which was useful for preventing potentially dangerous particles with diameter larger than 6 microns from entering the bloodstream when injecting contrast agents in ultrasounds. This study also made advances for environmental sustainability in microfluidics due to the lack of outside electronics driving the flow, which came as an advantage of using a gravity based device.

9.10.4 Two-Dimensional Chromatography

In some cases, the selectivity provided by the use of one column can be insufficient to provide resolution of analytes in complex samples. Two-dimensional chromatography aims to increase the resolution of these peaks by using a second column with different physico-chemical (chemical classification) properties. Since the mechanism of retention on this new solid support is different from the first dimensional separation, it can be possible to separate compounds by two-dimensional chromatography that are indistinguishable by one-dimensional chromatography. Furthermore, the separation on the second dimension occurs faster than the first dimension. An example of a TDC separation is where the sample is spotted at one corner of a square plate, developed, air-dried, then rotated by 90° and usually redeveloped in a second solvent system. Two-dimensional chromatography can be applied to GC or LC separations. The heart-cutting approach selects a specific region of interest on the first dimension for separation, and the comprehensive approach uses all analytes in the second-dimension separation.

9.10.5 Simulated Moving-Bed Chromatography

The simulated moving bed (SMB) technique is a variant of high performance liquid chromatography; it is used to separate particles and/or chemical compounds that would be difficult or impossible to resolve otherwise. This increased separation is brought about by a valve-and-column arrangement that is used to lengthen the stationary phase indefinitely. In the moving bed technique of preparative chromatography the feed entry and the analyte recovery are simultaneous and continuous, but because of practical difficulties with a continuously moving bed, simulated moving bed technique was proposed. In the simulated moving bed technique instead of moving the bed, the sample inlet and the analyte exit positions are moved continuously, giving the impression of a moving bed. True moving bed chromatography (TMBC) is only a theoretical concept. Its simulation, SMBC is achieved by the use of a multiplicity of columns in series and a complex valve arrangement. This valve arrangement provides for sample and solvent feed and analyte and waste takeoff at appropriate locations of any column, whereby it allows switching at regular intervals the sample entry in one direction, the solvent entry in the opposite direction, whilst changing the analyte and waste takeoff positions appropriately as well.

9.10.6 Pyrolysis Gas Chromatography

Pyrolysis–gas chromatography–mass spectrometry is a method of chemical analysis in which the sample is heated to decomposition to produce smaller molecules that are separated by gas chromatography and detected using mass spectrometry.

Pyrolysis is the thermal decomposition of materials in an inert atmosphere or a vacuum. The sample is put into direct contact with a platinum wire, or placed in a quartz sample tube, and rapidly heated to 600–1000 °C. Depending on the application even higher temperatures are used. Three different heating techniques are used in actual pyrolyzers: Isothermal furnace, inductive heating (Curie point filament), and resistive heating using platinum filaments. Large molecules cleave at their weakest points and produce smaller, more volatile fragments. These fragments can be separated by gas chromatography. Pyrolysis GC chromatograms are typically complex because a wide range of different decomposition products is formed. The data can either be used as fingerprints to prove material identity or the GC/MS data is used to identify individual fragments to obtain structural information. To increase the volatility of polar fragments, various methylating reagents can be added to a sample before pyrolysis.

Besides the usage of dedicated pyrolyzers, pyrolysis GC of solid and liquid samples can be performed directly inside Programmable Temperature Vaporizer (PTV) injectors that provide quick heating (up to 30 °C/s) and high maximum temperatures of 600–650 °C. This is sufficient for some pyrolysis applications. The main advantage is that no dedicated instrument has to be purchased and pyrolysis can be performed as part of routine GC analysis. In this case, quartz GC inlet liners have to be used. Quantitative data can be acquired, and good results of derivatization inside the PTV injector are published as well.

9.10.7 Fast Protein Liquid Chromatography

Fast protein liquid chromatography (FPLC), is a form of liquid chromatography that is often used to analyze or purify mixtures of proteins. As in other forms of chromatography, separation is possible because the different components of a mixture have different affinities for two materials, a moving fluid (the "mobile phase") and a porous solid (the stationary phase). In FPLC the mobile phase is an aqueous solution, or "buffer". The buffer flow rate is controlled by a positive-displacement pump and is normally kept constant, while the composition of the buffer can be varied by drawing fluids in different proportions from two or more external reservoirs. The stationary phase is a resin composed of beads, usually of cross-linked agarose, packed into a cylindrical glass or plastic column. FPLC resins are available in a wide range of bead sizes and surface ligands depending on the application.

9.10.8 Countercurrent Chromatography

Countercurrent chromatography (CCC) is a type of liquid-liquid chromatography, where both the stationary and mobile phases are liquids and the liquid stationary phase is held stagnant by a strong centrifugal force.

Hydrodynamic Countercurrent Chromatography (CCC)

The operating principle of CCC instrument requires a column consisting of an open tube coiled around a bobbin. The bobbin is rotated in a double-axis gyratory motion (a cardioid), which causes a variable gravity (G) field to act on the column during each rotation. This motion causes the column to see one partitioning step per revolution and components of the sample separate in the column due to their partitioning coefficient between the two immiscible liquid phases used. There are many types of CCC available today. These include HSCCC (High Speed CCC) and HPCCC (High Performance CCC). HPCCC is the latest and best-performing version of the instrumentation available currently.

Centrifugal Partition Chromatography (CPC)

In the CPC (centrifugal partition chromatography or hydrostatic countercurrent chromatography) instrument, the column consists of a series of cells interconnected by ducts attached to a rotor. This rotor rotates on its central axis creating the centrifugal field necessary to hold the stationary phase in place. The separation process in CPC is governed solely by the partitioning of solutes between the stationary and mobile phases, which mechanism can be easily described using the partition coefficients (K_D) of solutes. CPC instruments are commercially available for laboratory, pilot, and industrial-scale separations with different sizes of columns ranging from some 10 milliliters to 10 liters in volume.

9.10.9 Periodic Counter-Current Chromatography

In contrast to Counter current chromatography, periodic counter-current chromatography (PCC) uses a solid stationary phase and only a liquid mobile phase. It thus is much more similar to conventional affinity chromatography than to counter current chromatography. PCC uses multiple columns, which during the loading phase are connected in line. This mode allows for overloading the first column in this series without losing product, which already breaks through the column before the resin is fully saturated. The breakthrough product is captured on the subsequent column(s). In a next step the columns are disconnected from one another. The first column is washed and eluted, while the other column(s) are still being loaded. Once the (initially) first column is re-equilibrated, it is re-introduced to the loading stream, but as last column. The process then continues in a cyclic fashion.

9.10.10 Chiral Chromatography

Chiral chromatography involves the separation of stereoisomers. In the case of enantiomers, these have no chemical or physical differences apart from being three-dimensional mirror images. To enable chiral separations to take place, either the mobile phase or the stationary phase must themselves be made chiral, giving differing affinities between the analytes. Chiral chromatography HPLC columns (with a chiral stationary phase) in both normal and reversed phase are commercially available. Conventional chromatography are incapable of separating racemic mixtures of enantiomers. However, in some cases nonracemic mixtures of enantiomers may be separated unexpectedly by conventional liquid chromatography (e.g. HPLC without chiral mobile phase or stationary phase).

9.10.11 Aqueous Normal-Phase Chromatography

Aqueous normal-phase (ANP) chromatography is characterized by the elution behavior of classical normal phase mode (i.e. where the mobile phase is significantly less polar than the stationary phase) in which water is one of the mobile phase solvent system components. It is distinguished from hydrophilic interaction liquid chromatography (HILIC) in that the retention mechanism is due to adsorption rather than partitioning.

CHAPTER 10

Gel Filtration

Gel filtration, also known as size-exclusion chromatography (SEC), is a powerful and widely used technique in the field of bioseparations. It allows for the separation and purification of biomolecules based on their size and, to some extent, their shape. Unlike many other chromatographic methods, gel filtration does not rely on binding interactions between the stationary phase and the solute molecules. Instead, the separation is achieved through the differential migration of molecules through a porous gel matrix. This gentle and non-destructive method is particularly well-suited for proteins, nucleic acids, polysaccharides, and other macromolecules, making it a valuable tool in biochemical research, pharmaceutical manufacturing, and biotechnology.

Size-exclusion chromatography, also known as molecular sieve chromatography, is a chromatographic method in which molecules in solution are separated by their shape, and in some cases size. It is usually applied to large molecules or macromolecular complexes such as proteins and industrial polymers. Typically, when an aqueous solution is used to transport the sample through the column, the technique is known as gel filtration chromatography, versus the name gel permeation chromatography, which is used when an organic solvent is used as a mobile phase. The chromatography column is packed with fine, porous beads which are commonly composed of dextran, agarose, or polyacrylamide polymers. The pore sizes of these beads are used to estimate the dimensions of macromolecules. SEC is a widely used polymer characterization method because of its ability to provide good molar mass distribution (Mw) results for polymers.

Size-exclusion chromatography (SEC) is fundamentally different from all other chromatographic techniques in that separation is based on a simple procedure of classifying molecule sizes rather than any type of interaction.

10.1 APPLICATIONS

The main application of size-exclusion chromatography is the fractionation of proteins and other water-soluble polymers, while gel permeation chromatography is used to analyze the molecular weight distribution of organic-soluble polymers. Either technique should not be confused with gel electrophoresis, where an electric field is used to "pull" molecules through the gel depending on their electrical charges. The amount of time a solute remains within a pore is dependent on the size of the pore. Larger solutes will have access to a smaller volume and vice versa. Therefore, a smaller solute will remain within the pore for a longer period of time compared to a larger solute.

Even though size exclusion chromatography is widely utilized to study natural organic material, there are limitations. One of these limitations include that there is no standard molecular weight marker; thus, there is nothing to compare the results back to. If precise molecular weight is required, other methods should be used.

10.2 ADVANTAGES

The advantages of this method include good separation of large molecules from the small molecules with a minimal volume of eluate, and that various solutions can be applied without interfering with the filtration process, all while preserving the biological activity of the particles to separate. The technique is generally combined with others that further separate molecules by other characteristics, such as acidity, basicity, charge, and affinity for certain compounds. With size exclusion chromatography, there are short and well-defined separation times and narrow bands, which lead to good sensitivity. There is also no sample loss because solutes do not interact with the stationary phase.

The other advantage to this experimental method is that in certain cases, it is feasible to determine the approximate molecular weight of a compound. The shape and size of the compound (eluent) determine how the compound interacts with the gel (stationary phase). To determine approximate molecular weight, the elution volumes of compounds with their corresponding molecular weights are obtained and then a plot of "K_{av}" vs "log(Mw)" is made, where $K_{av} = (V_e - V_o)/(V_t - V_o)$ and Mw is the molecular mass. This plot acts as a calibration curve, which is used to approximate the desired compound's molecular weight. The V_e component represents the volume at which the intermediate molecules elute such as molecules that have partial access to the beads of the column. In addition, V_t is the sum of the total volume between the beads and the volume within the beads. The V_o component represents the volume at which the larger molecules elute, which elute in the beginning. Disadvantages are, for example, that only a limited number of bands can be accommodated because the time scale of the chromatogram is short, and, in general, there must be a 10% difference in molecular mass to have a good resolution.

10.3 DISCOVERY

The technique was invented in 1955 by Grant Henry Lathe and Colin R Ruthven, working at Queen Charlotte's Hospital, London. They later received the John Scott Award for this invention. While Lathe and Ruthven used starch gels as the matrix, Jerker Porath and Per Flodin later introduced dextran gels; other gels with size fractionation properties include agarose and polyacrylamide. A short review of these developments has appeared.

There were also attempts to fractionate synthetic high polymers; however, it was not until 1964, when J. C. Moore of the Dow Chemical Company published his work on the preparation of gel permeation chromatography (GPC) columns based on cross-linked polystyrene with controlled pore size, that a rapid increase of research activity in this field began. It was recognized almost immediately that with proper calibration, GPC was capable to provide molar mass and molar mass distribution information for synthetic polymers. Because the latter information was difficult to obtain by other methods, GPC came rapidly into extensive use.

10.4 THEORY AND METHOD

SEC is used primarily for the analysis of large molecules such as proteins or polymers. SEC works by trapping smaller molecules in the pores of the adsorbent ("stationary phase"). This process is usually performed within a column, which typically consists of a hollow tube tightly packed with micron-scale polymer beads containing pores of different sizes. These pores may be depressions on the surface or channels through the bead. As the solution travels down the column some particles enter into the pores. Larger particles cannot enter into as many pores. The larger the particles, the faster the elution. The larger molecules simply pass by the pores because those molecules are too large to enter the pores. Larger molecules therefore flow through the column more quickly than smaller molecules, that is, the smaller the molecule, the longer the retention time.

One requirement for SEC is that the analyte does not interact with the surface of the stationary phases, with differences in elution time between analytes ideally being based solely on the solute volume the analytes can enter, rather than chemical or electrostatic interactions with the stationary phases. Thus, a small molecule that can penetrate every region of the stationary phase pore system can enter a total volume equal to the sum of the entire pore volume and the interparticle volume. This small molecule elutes late (after the molecule has penetrated all of the pore- and interparticle volume—approximately 80% of the column volume). At the other extreme, a very large molecule that cannot penetrate any the smaller pores can enter only the interparticle volume (~35% of the column volume) and elutes earlier when this volume of mobile phase has passed through the column. The underlying principle of SEC is that particles of different sizes elute (filter) through a stationary phase at different rates. This results in the separation of a solution of particles based on size. Provided that all the particles are loaded simultaneously or near-simultaneously, particles of the same size should elute together.

Fig: Agarose-Based SEC Columns Used for Protein Purification on an AKTA FPLC Machine

However, as there are various measures of the size of a macromolecule (for instance, the radius of gyration and the hydrodynamic radius), a fundamental problem in the theory of SEC has been the choice of a proper molecular size parameter by which molecules of different kinds are separated. Experimentally, Benoit and co-workers found an excellent correlation between elution volume and a dynamically based molecular size, the hydrodynamic volume, for several different chain architecture and chemical compositions. The observed correlation based on the hydrodynamic volume became accepted as the basis of universal SEC calibration.

Still, the use of the hydrodynamic volume, a size based on dynamical properties, in the interpretation of SEC data is not fully understood. This is because SEC is typically run under low flow rate conditions where hydrodynamic factor should have little effect on the separation. In fact, both theory and computer simulations assume a thermodynamic separation principle: the separation process is determined by the equilibrium distribution (partitioning) of solute macromolecules between two phases: a dilute bulk solution phase located at the interstitial space and confined solution phases within the pores of column packing material. Based on this theory, it has been shown that the relevant size parameter to the partitioning of polymers in pores is the mean span dimension (mean maximal projection onto a line). Although this issue has not been fully resolved, it is likely that the mean span dimension and the hydrodynamic volume are strongly correlated.

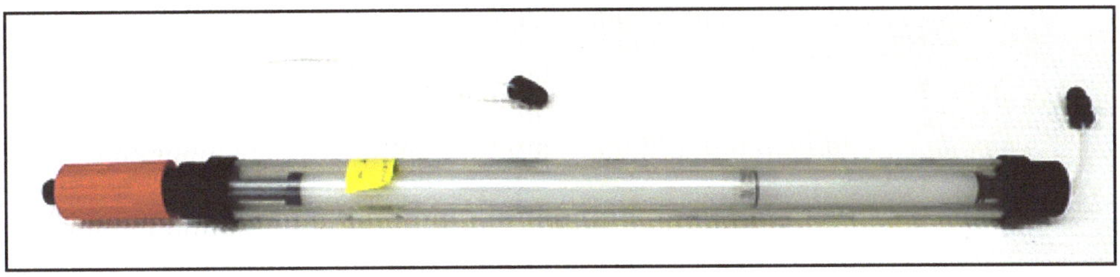

Fig: A Size Exclusion Column

Each size exclusion column has a range of molecular weights that can be separated. The exclusion limit defines the molecular weight at the upper end of the column 'working' range and is where molecules are too large to get trapped in the stationary phase. The lower end of the range is defined by the permeation limit, which defines the molecular weight of a molecule that is small enough to penetrate all pores of the stationary phase. All molecules below this molecular mass are so small that they elute as a single band.

The filtered solution that is collected at the end is known as the eluate. The void volume includes any particles too large to enter the medium, and the solvent volume is known as the column volume.

Following are the materials which are commonly used for porous gel beads in size exclusion chromatography

Sr. No	Material And Trade Name	Fractionation Range (kDa)
1	Sephadex G-10	0-0.7
2	Sephadex G-25	1-5
3	Sephadex G-50	1.5-30
4	Sephadex G-75	3-70
5	Sephadex G-100	4-150

6	Sephadex G-150	5-300
7	Sephadex G-200	5-8000
8	Bio-gel P-2	0.1-1.8
9	Bio-gel P-6	1-6
10	Bio-gel P-60	3-60
11	Bio-gel P-150	1.5-150
12	Bio-gel P-300	16-400
13	Sepharose 2B	2000-25000
14	Sepharose 4B	300-3000
15	Sepharose 6B	10-20000

10.4.1 Factors Affecting Filtration

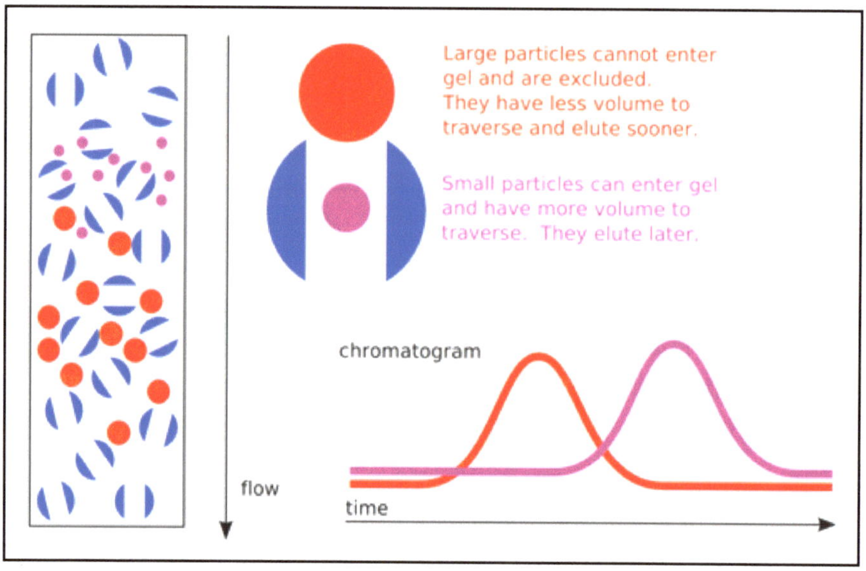

Fig: A Cartoon Illustrating the Theory behind Size Exclusion Chromatography

In real-life situations, particles in solution do not have a fixed size, resulting in the probability that a particle that would otherwise be hampered by a pore passing right by it. Also, the stationary-phase particles are not ideally defined; both particles and pores may vary in size. Elution curves, therefore, resemble Gaussian distributions. The stationary phase may also interact in undesirable ways with a particle and influence retention times, though great care is taken by column manufacturers to use stationary phases that are inert and minimize this issue.

Like other forms of chromatography, increasing the column length enhances resolution, and increasing the column diameter increases column capacity. Proper column packing is important for maximum resolution: An over-packed column can collapse the pores in the beads, resulting in a loss of resolution. An under-packed column can reduce the relative surface area of the stationary phase accessible to smaller species, resulting in those species spending less time trapped in pores. Unlike affinity chromatography techniques, a solvent head at the top of the column can drastically diminish resolution as the sample diffuses prior to loading, broadening the downstream elution.

10.4.2 Analysis

In simple manual columns, the eluent is collected in constant volumes, known as fractions. The more similar the particles are in size the more likely they are in the same fraction and not detected separately. More advanced columns overcome this problem by constantly monitoring the eluent.

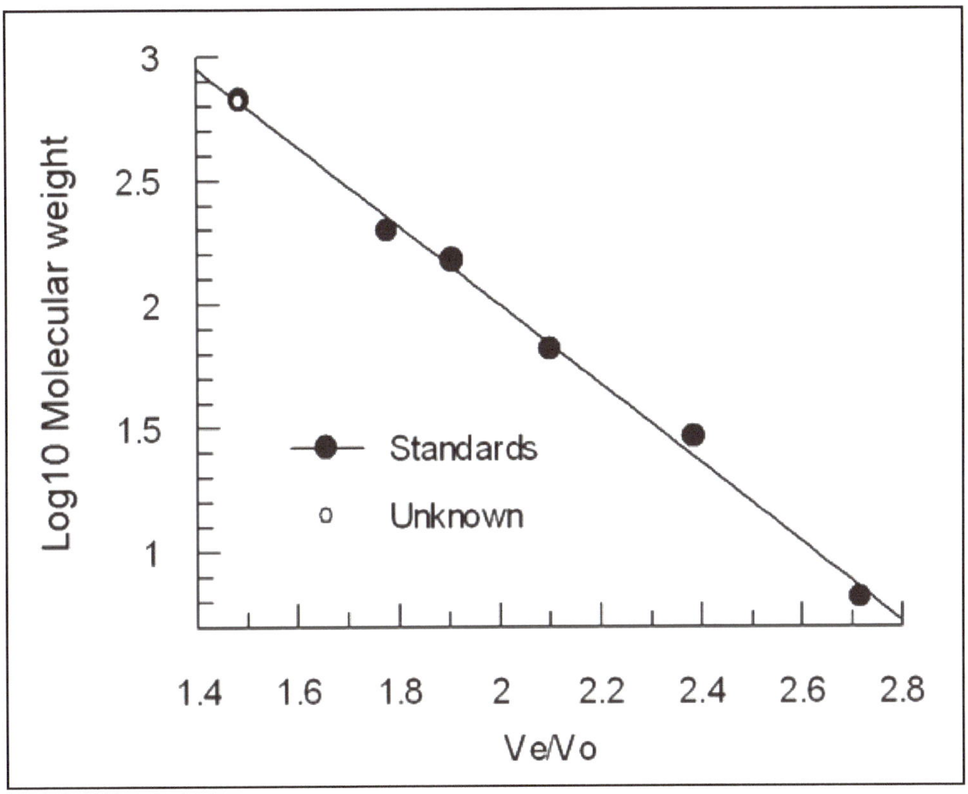

Fig: Standardization (Calibration) of a Size Exclusion Column

The collected fractions are often examined by spectroscopic techniques to determine the concentration of the particles eluted. Common spectroscopy detection techniques are refractive index (RI) and ultraviolet (UV). When eluting spectroscopically similar species (such as during biological purification), other techniques may be necessary to identify the contents of each fraction. It is also possible to analyze the eluent flow continuously with RI, LALLS, Multi-Angle Laser Light Scattering MALS, UV, and/or viscosity measurements.

The elution volume (Ve) decreases roughly linear with the logarithm of the molecular hydrodynamic volume. Columns are often calibrated using 4-5 standard samples (e.g., folded proteins of known molecular weight), and a sample containing a very large molecule such as thyroglobulin to determine the void volume. (Blue dextran is not recommended for Vo determination because it is heterogeneous and may give variable results) The elution volumes of the standards are divided by the elution volume of the thyroglobulin (Ve/Vo) and plotted against the log of the standards' molecular weights.

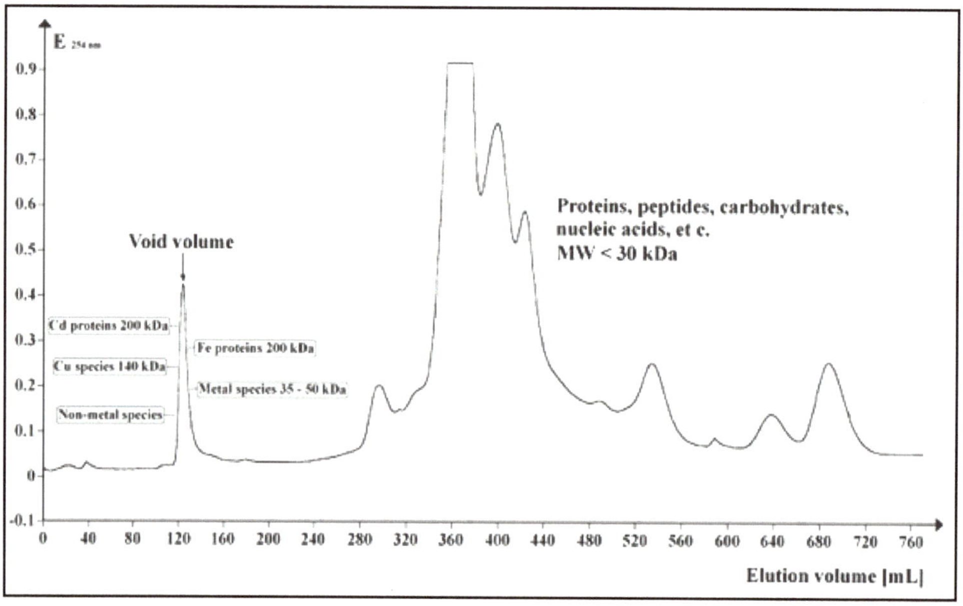

Fig: Size Exclusion Chromatogram after Bioanalytical Continuous-Elution Gel Chromatography of a Plant Sample

10.5 APPLICATIONS

10.5.1 Biochemical Applications

In general, SEC is considered a low-resolution chromatography as it does not discern similar species very well, and is therefore often reserved for the final step of a purification.

The technique can determine the quaternary structure of purified proteins that have slow exchange times, since it can be carried out under native solution conditions, preserving macromolecular interactions. SEC can also assay protein tertiary structure, as it measures the hydrodynamic volume (not molecular weight), allowing folded and unfolded versions of the same protein to be distinguished. For example, the apparent hydrodynamic radius of a typical protein domain might be 14 Å and 36 Å for the folded and unfolded forms, respectively.

SEC allows the separation of these two forms, as the folded form elutes much later due to its smaller size.

10.5.2 Polymer Synthesis

SEC can be used as a measure of both the size and the polydispersity of a synthesized polymer, that is, the ability to find the distribution of the sizes of polymer molecules. If standards of a known size are run previously, then a calibration curve can be created to determine the sizes of polymer molecules of interest in the solvent chosen for analysis (often THF). In alternative fashion, techniques such as light scattering and/or viscometry can be used online with SEC to yield absolute molecular weights that do not rely on calibration with standards of known molecular weight. Due to the difference in size of two polymers with identical molecular weights, the absolute determination methods are, in general, more desirable. A typical SEC system can quickly (in about half an hour) give polymer chemists information on the size and polydispersity of the sample. The preparative SEC can be used for polymer fractionation on an analytical scale.

10.6 DRAWBACKS

In SEC, mass is not measured so much as the hydrodynamic volume of the polymer molecules, that is, how much space a particular polymer molecule takes up when it is in solution. However, the approximate molecular weight can be calculated from SEC data because the exact relationship between molecular weight and hydrodynamic volume for polystyrene can be found. For this, polystyrene is used as a standard. But the relationship between hydrodynamic volume and molecular weight is not the same for all polymers, so only an approximate measurement can be obtained. Another drawback is the possibility of interaction between the stationary phase and the analyte. Any interaction leads to a later elution time and thus mimics a smaller analyte size.

When performing this method, the bands of the eluting molecules may be broadened. This can occur by turbulence caused by the flow of the mobile phase molecules passing through the molecules of the stationary phase. In addition, molecular thermal diffusion and friction between the molecules of the glass walls and the molecules of the eluent contribute to the broadening of the bands. Besides broadening, the bands also overlap with each other. As a result, the eluent usually gets considerably diluted. A few precautions can be taken to prevent the likelihood of the bands broadening. For instance, one can apply the sample in a narrow, highly concentrated band on the top of the column. The more concentrated the eluent is, the more efficient the procedure would be. However, it is not always possible to concentrate the eluent, which can be considered as one more disadvantage.

10.7 ABSOLUTE SIZE-EXCLUSION CHROMATOGRAPHY

Absolute size-exclusion chromatography (ASEC) is a technique that couples a light scattering instrument, most commonly multi-angle light scattering (MALS) or another form of static light scattering (SLS), but possibly a dynamic light scattering (DLS) instrument, to a size-exclusion chromatography system for absolute molar mass and/or size measurements of proteins and macromolecules as they elute from the chromatography system.

The definition of "absolute" in this case is that calibration of retention time on the column with a set of reference standards is not required to obtain molar mass or the hydrodynamic size, often referred to as hydrodynamic diameter (D_H in units of nm). Non-ideal column interactions, such as electrostatic or hydrophobic surface interactions that modulate retention time relative to standards, do not impact the final result.

Likewise, differences between conformation of the analyte and the standard have no effect on an absolute measurement; for example, with MALS analysis, the molar mass of inherently disordered proteins are characterized accurately even though they elute at much earlier times than globular proteins with the same molar mass, and the same is true of branched polymers which elute late compared to linear reference standards with the same molar mass. Another benefit of ASEC is that the molar mass and/or size is determined at each point in an eluting peak, and therefore indicates homogeneity or polydispersity within the peak.

For example, SEC-MALS analysis of a monodisperse protein will show that the entire peak consists of molecules with the same molar mass, something that is not possible with standard SEC analysis.

Determination of molar mass with SLS requires combining the light scattering measurements with concentration measurements. Therefore SEC-MALS typically includes the light scattering detector and either a differential refractometer or UV/Vis absorbance detector. In addition, MALS determines the rms radius R_g of molecules above a certain size limit, typically 10 nm. SEC-MALS can therefore analyze the conformation of polymers via the relationship of molar mass to R_g. For smaller molecules, either DLS or, more commonly, a differential viscometer is added to determine hydrodynamic radius and evaluate molecular conformation in the same manner.

In SEC-DLS, the sizes of the macromolecules are measured as they elute into the flow cell of the DLS instrument from the size exclusion column set. The hydrodynamic size of the molecules or particles are measured and not their molecular weights. For proteins a Mark-Houwink type of calculation can be used to estimate the molecular weight from the hydrodynamic size.

A major advantage of DLS coupled with SEC is the ability to obtain enhanced DLS resolution. Batch DLS is quick and simple and provides a direct measure of the average size, but the baseline resolution of DLS is a ratio of 3:1 in diameter. Using SEC, the proteins and protein oligomers are separated, allowing oligomeric resolution. Aggregation studies can also be done using ASEC. Though the aggregate concentration may not be calculated with light scattering (an online concentration detector such as that used in SEC-MALS for molar mass measurement also determines aggregate concentration), the size of the aggregate can be measured, only limited by the maximum size eluting from the SEC columns.

Limitations of ASEC with DLS detection include flow-rate, concentration, and precision. Because a correlation function requires anywhere from 3–7 seconds to properly build, a limited number of data points can be collected across the peak. ASEC with SLS detection is not limited by flow rate and measurement time is essentially instantaneous, and the range of concentration is several orders of magnitude larger than for DLS. However, molar mass analysis with SEC-MALS does require accurate concentration measurements. MALS and DLS detectors are often combined in a single instrument for more comprehensive absolute analysis following separation by SEC.

10.8 KEY PARAMETERS AFFECTING SEPARATION

Several important parameters influence the efficiency and resolution of gel filtration chromatography. One of the most significant is the pore size of the gel, which defines the fractionation range of the column. Molecules that are much larger than the pores are completely excluded and elute at the void volume, while very small molecules may enter almost all the pores and are eluted later, near the total column volume. Between these extremes, molecules are separated in order of decreasing size.

Column dimensions also play a role in separation performance. Longer columns provide better resolution because they allow more time for differential migration, but they also require longer run times and higher pressures. Column diameter determines the sample volume that can be loaded without loss of resolution. Overloading the column with too much sample can result in peak broadening and decreased separation efficiency.

Flow rate is another important factor. Lower flow rates generally improve resolution by allowing molecules more time to partition between the mobile and stationary phases. However, very slow flow rates can be impractical for large-scale applications. Therefore, an optimal balance must be achieved between resolution and throughput.

Buffer composition is important not only for maintaining the native structure and function of the biomolecules but also for preventing non-specific interactions with the matrix. A neutral pH and moderate ionic strength are typically used to reduce electrostatic effects and ensure consistent flow characteristics.

10.9 APPLICATIONS OF GEL FILTRATION IN BIOSEPARATIONS

Gel filtration chromatography has a broad range of applications in the separation and purification of biomolecules. One of its primary uses is in the purification of proteins and peptides, particularly in cases where a gentle, non-denaturing method is required. Because the technique does not involve any binding interactions or harsh conditions, it is ideal for preserving the biological activity of enzymes, antibodies, and other sensitive proteins. It is often employed as a polishing step in downstream processing, following ion exchange or affinity chromatography, to remove aggregates, buffer components, or contaminants of different sizes.

In molecular biology, gel filtration is frequently used to desalt samples or to exchange buffers. It is effective for removing low-molecular-weight salts, detergents, or small organic molecules from protein or nucleic acid preparations. This is especially useful before applications such as electrophoresis, crystallization, or enzyme assays where specific buffer conditions are required.

Another important application is in molecular weight determination. By calibrating the column with standards of known molecular weight, researchers can estimate the size of unknown proteins or complexes based on their elution volume. This has been instrumental in studying protein oligomerization, conformational states, and the formation of protein complexes.

Gel filtration is also used in the analysis of biopharmaceuticals, particularly in the detection and quantification of aggregates, which can affect drug efficacy and safety. It plays a crucial role in quality control during the production of monoclonal antibodies and other therapeutic proteins.

10.10 ADVANTAGES OF GEL FILTRATION CHROMATOGRAPHY

One of the main advantages of gel filtration is its gentle and non-destructive nature. Because the separation is based purely on size and does not involve any chemical or physical interaction with the stationary phase, the integrity and activity of delicate biomolecules are preserved. This makes it suitable for labile proteins, enzymes, and complexes that might be denatured by harsher methods.

The method also allows for simultaneous separation and buffer exchange, which can be advantageous in sample preparation workflows. For example, a protein can be both purified from contaminants and transferred into a new buffer in a single step. This versatility simplifies downstream processes and reduces the need for additional purification steps.

Gel filtration provides high reproducibility and is relatively easy to optimize and scale. It requires minimal sample preparation and can be automated using modern liquid chromatography systems. Columns can be reused multiple times with proper maintenance, contributing to cost-effectiveness over time.

10.11 LIMITATIONS AND CHALLENGES

Despite its many benefits, gel filtration does have limitations. Its resolution is generally lower than that of other chromatographic techniques such as ion exchange or affinity chromatography. This is because separation is limited to differences in hydrodynamic volume, which may not always be sufficient to distinguish closely related molecules.

The sample loading capacity is also relatively low. Overloading the column leads to loss of resolution and poor separation, making gel filtration less suitable as the primary step for large-scale purifications. Instead, it is best used as a final polishing step or for analytical purposes.

Another limitation is the requirement for relatively large volumes of buffer and longer run times, especially for larger columns or higher-resolution applications. This can lead to increased processing time and resource consumption, particularly in large-scale operations.

Furthermore, the technique is not suitable for separating molecules of similar size but different properties, such as charge or hydrophobicity. In such cases, other chromatographic methods may be more appropriate.

10.12 RECENT ADVANCES AND INNOVATIONS

Recent developments in gel filtration technology aim to overcome some of these limitations and expand its utility. High-performance gel filtration columns with smaller particle sizes and narrower pore distributions have been introduced to improve resolution. These columns offer better separation of complex mixtures and are particularly valuable in analytical applications such as protein characterization and aggregate detection.

Miniaturized and microfluidic versions of gel filtration are being developed for lab-on-a-chip applications. These devices offer faster analysis with lower sample and reagent consumption, making them ideal for high-throughput screening and point-of-care diagnostics.

Integration with multi-angle light scattering (MALS) and other detection systems has enhanced the analytical power of gel filtration, enabling precise determination of molecular weight, size, and oligomeric state in real time. These innovations are particularly useful in biopharmaceutical development and structural biology.

10.13 CONCLUSION

Gel filtration remains a cornerstone technique in bioseparations due to its simplicity, versatility, and non-denaturing properties. It is especially valuable for the purification and characterization of proteins, nucleic acids, and other macromolecules where maintaining structural integrity is essential. While it may not offer the highest resolution or loading capacity compared to other methods, its unique advantages make it indispensable in many laboratory and industrial settings. With continued advancements in column technology and instrumentation, gel filtration is poised to remain a vital component of the bioseparation toolbox for years to come.

CHAPTER 11

Ion Exchange Chromatography and Chromatofocusing

11.1 ION EXCHANGE CHROMATOGRAPHY

Ion chromatography (or ion-exchange chromatography) is a form of chromatography that separates ions and ionizable polar molecules based on their affinity to the ion exchanger. It works on almost any kind of charged molecule—including small inorganic anions, large proteins, small nucleotides, and amino acids. However, ion chromatography must be done in conditions that are one pH unit away from the isoelectric point of a protein.

The two types of ion chromatography are anion-exchange and cation-exchange. Cation-exchange chromatography is used when the molecule of interest is positively charged. The molecule is positively charged because the pH for chromatography is less than the pI (also known as pH(I)). In this type of chromatography, the stationary phase is negatively charged and positively charged molecules are loaded to be attracted to it. Anion-exchange chromatography is when the stationary phase is positively charged and negatively charged molecules (meaning that pH for chromatography is greater than the pI) are loaded to be attracted to it. It is often used in protein purification, water analysis, and quality control. The water-soluble and charged molecules such as proteins, amino acids, and peptides bind to moieties which are oppositely charged by forming ionic bonds to the insoluble stationary phase.

The equilibrated stationary phase consists of an ionizable functional group where the targeted molecules of a mixture to be separated and quantified can bind while passing through the column—a cationic stationary phase is used to separate anions and an anionic stationary phase is used to separate cations. Cation exchange chromatography is used when the desired molecules to separate are cations and anion exchange chromatography is used to separate anions. The bound molecules then can be eluted and collected using an eluant which contains anions and cations by running a higher concentration of ions through the column or by changing the pH of the column.

One of the primary advantages for the use of ion chromatography is that only one interaction is involved the separation, as opposed to other separation techniques; therefore, ion chromatography may have higher matrix tolerance. Another advantage of ion exchange is the predictability of elution patterns (based on the presence of the ionizable group). For example, when cation exchange chromatography is used, certain cations will elute out first and others later. A local charge balance is always maintained. However, there are also disadvantages involved when performing ion-exchange chromatography, such as constant evolution of the technique which leads to the inconsistency from column to column. A major limitation to this purification technique is that it is limited to ionizable group.

11.1.1 History

Ion chromatography has advanced through the accumulation of knowledge over a course of many years. Starting from 1947, Spedding and Powell used displacement ion-exchange chromatography for the separation of the rare earths. Additionally, they showed the ion-exchange separation of 14N and 15N isotopes in ammonia. At the start of the 1950s, Kraus and Nelson demonstrated the use of many analytical methods for metal ions dependent on their separation of their chloride, fluoride, nitrate or sulfate complexes by anion chromatography. Automatic in-line detection was progressively introduced from 1960 to 1980 as well as novel chromatographic methods for metal ion separations. A groundbreaking method by Small, Stevens and Bauman at Dow Chemical Co. unfolded the creation of the modern ion chromatography. Anions and cations could now be separated efficiently by a system of suppressed conductivity detection. In 1979, a method for anion chromatography with non-suppressed conductivity detection was introduced by Gjerde et al. Following it in 1980, was a similar method for cation chromatography.

ca. 1850	Soil as an ion exchanger for Mg^{2+}, Ca^{2+} and NH_4^+	Thomson & Way
1935	Sulfonated and aminated condensation polymers (phenol/formaldehyde)	Adams, Homes
1942	Sulfonated PS/DVB resin as cation exchanger (Manhattan Project)	d'Alelio
1947	Aminated PS/DVB resin as anion exchanger	McBurney
1953	Ion exclusion chromatography	Wheaton, Baumann
1957	Macroporous ion exchangers	Corte, Meyer, Kunin et al.
1959	Basic Theoretical principles	Helfferich
1967-70	Pellicular ion exchnagers	Horvath, Kirkland
1975	Ion exchange chromatography with conductivity detection using a "stripper"	Small, Stevens, Bauann
1979	Conductivity detection without a "stripper"	Gjerde, Fritz, Schmuckler
1976-80	Ion pair chromatography	Waters, Bidlingmeier, Horvath et al.

Fig: Ion Exchange Chromatography

As a result, a period of extreme competition began within the IC market, with supporters for both suppressed and non-suppressed conductivity detection. This competition led to fast growth of new forms and the fast evolution of IC. A challenge that needs to be overcome in the future development of IC is the preparation of highly efficient monolithic ion-exchange columns and overcoming this challenge would be of great importance to the development of IC.

The boom of Ion exchange chromatography primarily began between 1935 and 1950 during World War II and it was through the "Manhattan project" that applications and IC were significantly extended. Ion chromatography was originally introduced by two English researchers, agricultural Sir Thompson and chemist J T Way. The works of Thompson and Way involved the action of water-soluble fertilizer salts, ammonium sulfate and potassium chloride. These salts could not easily be extracted from the ground due to the rain. They performed ion methods to treat clays with the salts, resulting in the extraction of ammonia in addition to the release of calcium. It was in the fifties and sixties that theoretical models were developed for IC for further understanding and it was not until the seventies that continuous detectors were utilized, paving the path for the development from low-pressure to high-performance chromatography. Not until 1975 was "ion chromatography" established as a name in reference to the techniques, and was thereafter used as a name for marketing purposes. Today IC is important for investigating aqueous systems, such as drinking water. It is a popular method for analyzing anionic elements or complexes that help solve environmentally relevant problems. Likewise, it also has great uses in the semiconductor industry.

Because of the abundant separating columns, elution systems, and detectors available, chromatography has developed into the main method for ion analysis.

When this technique was initially developed, it was primarily used for water treatment. Since 1935, ion exchange chromatography rapidly manifested into one of the most heavily leveraged techniques, with its principles often being applied to majority of fields of chemistry, including distillation, adsorption, and filtration.

11.1.2 Principle

Ion-exchange chromatography separates molecules based on their respective charged groups. Ion-exchange chromatography retains analyte molecules on the column based on coulombic (ionic) interactions. The ion exchange chromatography matrix consists of positively and negatively charged ions. Essentially, molecules undergo electrostatic interactions with opposite charges on the stationary phase matrix. The stationary phase consists of an immobile matrix that contains charged ionizable functional groups or ligands. The stationary phase surface displays ionic functional groups (R-X) that interact with analyte ions of opposite charge. To achieve electroneutrality, these immobilized charges couple with exchangeable counterions in the solution. Ionizable molecules that are to be purified, compete with these exchangeable counterions, for binding to the immobilized charges on the stationary phase. These ionizable molecules are retained or eluted based on their charge. Initially, molecules that do not bind or bind weakly to the stationary phase are first to be washed away. Altered conditions are needed for the elution of the molecules that bind to the stationary phase. The concentration of the exchangeable counterions, which competes with the molecules for binding, can be increased, or the pH can be changed to affect the ionic charge of the eluent or the solute. A change in pH affects the charge on the particular molecules and, therefore, alter their binding. When reducing the net charge of the solute's molecules, they start eluting out. This way, such adjustments can be used to release the proteins of interest. Additionally, concentration of counterions can be gradually varied to affect the retention of the ionized molecules, thus separate them. This type of elution is called gradient elution. On the other hand, step elution can be used, in which the concentration of counterions are varied in steps. This type of chromatography is further subdivided into cation exchange chromatography and anion-exchange chromatography. Positively charged molecules bind to cation exchange resins, while negatively charged molecules bind to anion exchange resins. The ionic compound consisting of the cationic species M+ and the anionic species B- can be retained by the stationary phase.

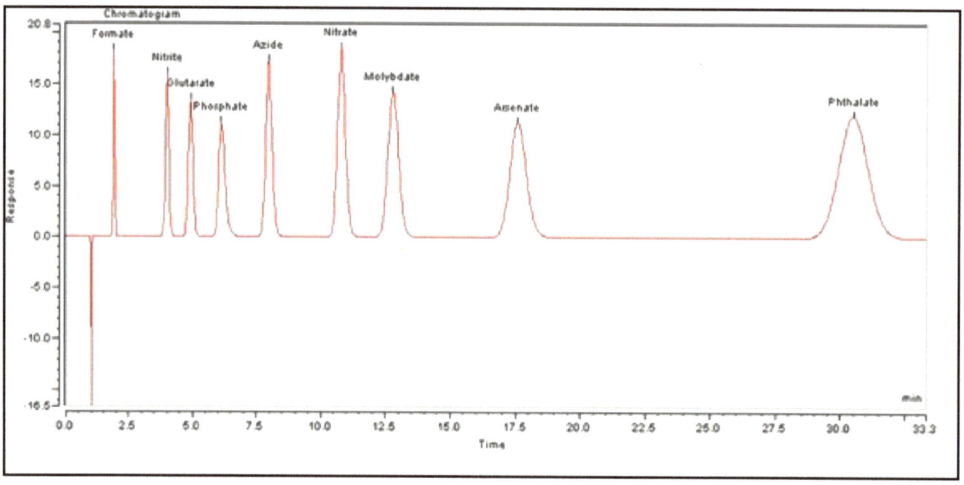

Fig: Ion Chromatogram Displaying Anion Separation

Cation exchange chromatography retains positively charged cations because the stationary phase displays a negatively charged functional group:

$$R\text{-}X^-C^+ + M^+B^- \rightleftarrows R\text{-}X^-M^+ + C^+ + B^-$$

Anion exchange chromatography retains anions using positively charged functional group:

$$R\text{-}X^+A^- + M^+B^- \rightleftarrows R\text{-}X^+B^- + M^+ + A^-$$

Note that the ion strength of either C+ or A- in the mobile phase can be adjusted to shift the equilibrium position, thus retention time.

The ion chromatogram shows a typical chromatogram obtained with an anion exchange column.

11.1.3 Procedure

Before ion-exchange chromatography can be initiated, it must be equilibrated. The stationary phase must be equilibrated to certain requirements that depend on the experiment that you are working with. Once equilibrated, the charged ions in the stationary phase will be attached to its opposite charged exchangeable ions, such as Cl^- or Na^+. Next, a buffer should be chosen in which the desired protein can bind to. After equilibration, the column needs to be washed. The washing phase will help elute out all impurities that does not bind to the matrix while the protein of interest remains bounded. This sample buffer needs to have the same pH as the buffer used for equilibration to help bind the desired proteins. Uncharged proteins will be eluted out of the column at a similar speed of the buffer flowing through the column with no retention. Once the sample has been loaded onto to the column, and the column has been washed with the buffer to elute out all non-desired proteins, elution is carried out at specific conditions to elute the desired proteins that are bound to the matrix. Bound proteins are eluted out by utilizing a gradient of linearly increasing salt concentration. With increasing ionic strength of the buffer, the salt ions will compete with the desired proteins in order to bind to charged groups on the surface of the medium. This will cause desired proteins to be eluted out of the column. Proteins that have a low net charge will be eluted out first as the salt concentration increases causing the ionic strength to increase. Proteins with high net charge will need a higher ionic strength for them to be eluted out of the column.

It is possible to perform ion exchange chromatography in bulk, on thin layers of medium such as glass or plastic plates coated with a layer of the desired stationary phase, or in chromatography columns. Thin layer chromatography or column chromatography share similarities in that they both act within the same governing principles; there is constant and frequent exchange of molecules as the mobile phase travels along the stationary phase. It is not imperative to add the sample in minute volumes as the predetermined conditions for the exchange column have been chosen so that there will be strong interaction between the mobile and stationary phases. Furthermore, the mechanism of the elution process will cause a compartmentalization of the differing molecules based on their respective chemical characteristics. This phenomenon is due to an increase in salt concentrations at or near the top of the column, thereby displacing the molecules at that position, while molecules bound lower are released at a later point when the higher salt concentration reaches that area. These principles are the reasons that ion exchange chromatography is an excellent candidate for initial chromatography steps in a complex purification procedure as it can quickly yield small volumes of target molecules regardless of a greater starting volume.

Comparatively simple devices are often used to apply counterions of increasing gradient to a chromatography column. Counterions such as copper (II) are chosen most often for effectively separating peptides and amino acids through complex formation.

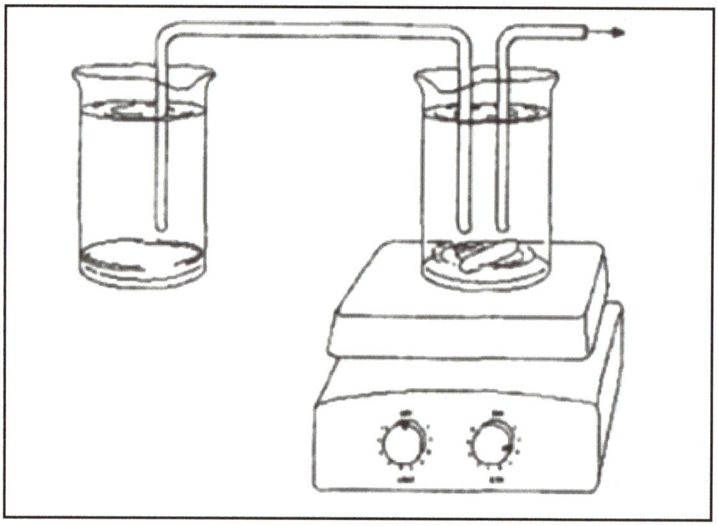

Fig: Chamber (Left) Contains High Salt Concentration. Stirred Chamber (Right) Contains Low Salt Concentration. Gradual Stirring Causes the Formation of a Salt Gradient as Salt Travel from High to Low Concentrations

A simple device can be used to create a salt gradient. Elution buffer is consistently being drawn from the chamber into the mixing chamber, thereby altering its buffer concentration. Generally, the buffer placed into the chamber is usually of high initial concentration, whereas the buffer placed into the stirred chamber is usually of low concentration. As the high concentration buffer from the left chamber is mixed and drawn into the column, the buffer concentration of the stirred column gradually increase. Altering the shapes of the stirred chamber, as well as of the limit buffer, allows for the production of concave, linear, or convex gradients of counterion.

A multitude of different mediums are used for the stationary phase. Among the most common immobilized charged groups used are trimethylaminoethyl (TAM), triethylaminoethyl (TEAE), diethyl-2-hydroxypropylaminoethyl (QAE), aminoethyl (AE), diethylaminoethyl (DEAE), sulpho (S), sulphomethyl (SM), sulphopropyl (SP), carboxy (C), and carboxymethyl (CM).

Successful packing of the column is an important aspect of ion chromatography. Stability and efficiency of a final column depends on packing methods, solvent used, and factors that affect mechanical properties of the column. In contrast to early inefficient dry- packing methods, wet slurry packing, in which particles that are suspended in an appropriate solvent are delivered into a column under pressure, shows significant improvement. Three different approaches can be employed in performing wet slurry packing: the balanced density method (solvent's density is about that of porous silica particles), the high viscosity method (a solvent of high viscosity is used), and the low viscosity slurry method (performed with low viscosity solvents).

Polystyrene is used as a medium for ion- exchange. It is made from the polymerization of styrene with the use of divinylbenzene and benzoyl peroxide. Such exchangers form hydrophobic interactions with proteins which can be irreversible. Due to this property, polystyrene ion exchangers are not suitable for protein separation. They are used on the other hand for the separation of small molecules in amino acid separation and removal of salt from water. Polystyrene ion exchangers with large pores can be used for the separation of protein but must be coated with a hydrophilic substance.

Cellulose based medium can be used for the separation of large molecules as they contain large pores. Protein binding in this medium is high and has low hydrophobic character. DEAE is an anion exchange matrix that is produced from a positive side group of diethylaminoethyl bound to cellulose or Sephadex.

Agarose gel based medium contain large pores as well but their substitution ability is lower in comparison to dextrans. The ability of the medium to swell in liquid is based on the cross-linking of these substances, the pH and the ion concentrations of the buffers used.

Incorporation of high temperature and pressure allows a significant increase in the efficiency of ion chromatography, along with a decrease in time. Temperature has an influence of selectivity due to its effects on retention properties. The retention factor ($k = (t_R^g - t_M^g)/(t_M^g - t_{ext})$) increases with temperature for small ions, and the opposite trend is observed for larger ions.

Despite ion selectivity in different mediums, further research is being done to perform ion exchange chromatography through the range of 40–175 °C.

An appropriate solvent can be chosen based on observations of how column particles behave in a solvent. Using an optical microscope, one can easily distinguish a desirable dispersed state of slurry from aggregated particles.

11.1.4 Weak and Strong Ion Exchangers

A "strong" ion exchanger will not lose the charge on its matrix once the column is equilibrated and so a wide range of pH buffers can be used. "Weak" ion exchangers have a range of pH values in which they will maintain their charge. If the pH of the buffer used for a weak ion exchange column goes out of the capacity range of the matrix, the column will lose its charge distribution and the molecule of interest may be lost. Despite the smaller pH range of weak ion exchangers, they are often used over strong ion exchangers due to their having greater specificity. In some experiments, the retention times of weak ion exchangers are just long enough to obtain desired data at a high specificity.

Resins (often termed 'beads') of ion exchange columns may include functional groups such as weak/strong acids and weak/strong bases. There are also special columns that have resins with amphoteric functional groups that can exchange both cations and anions. Some examples of functional groups of strong ion exchange resins are quaternary ammonium cation (Q), which is an anion exchanger, and sulfonic acid (S, $-SO_2OH$), which is a cation exchanger. These types of exchangers can maintain their charge density over a pH range of 0–14. Examples of functional groups of Weak ion exchange resins include diethylaminoethyl (DEAE, $-C_2H_4N(C_2H_5)_2$), which is an anion exchanger, and carboxymethyl (CM, $-CH_2-COOH$), which is a cation exchanger. These two types of exchangers can maintain the charge density of their columns over a pH range of 5–9.

Table: The Commonly Used Ion Exchangers

Sr. No	Name	Type	Functional group
1	DEAE Cellulose (Anion exchanger)	Weakly basic	DEAE (Diethylaminoethyl)
2	QAE Sephadex (Anion exchanger)	Strongly basic	QAE (Quaternary aminoethyl)
3	Q Sepharose (Anion exchanger)	Strongly basic	Q (Quaternary ammonium)
4	CM- Cellulose (Cation exchanger)	Weakly acidic	CM (Carboxymethyl)
5	SP Sepharose (Cation exchanger)	Strongly acidic	SP (Sulfopropyl)
6	SOURCE S (Cation exchanger)	Strongly acidic	S (Methyl sulfate)

In ion chromatography, the interaction of the solute ions and the stationary phase based on their charges determines which ions will bind and to what degree. When the stationary phase features positive groups which attracts anions, it is called an anion exchanger; when there are negative groups on the stationary phase, cations are attracted and it is a cation exchanger. The attraction between ions and stationary phase also depends on the resin, organic particles used as ion exchangers.

Each resin features relative selectivity which varies based on the solute ions present who will compete to bind to the resin group on the stationary phase. The selectivity coefficient, the equivalent to the equilibrium constant, is determined via a ratio of the concentrations between the resin and each ion, however, the general trend is that ion exchangers prefer binding to the ion with a higher charge, smaller hydrated radius, and higher polarizability, or the ability for the electron cloud of an ion to be disrupted by other charges. Despite this selectivity, excess amounts of an ion with a lower selectivity introduced to the column would cause the lesser ion to bind more to the stationary phase as the selectivity coefficient allows fluctuations in the binding reaction that takes place during ion exchange chromatography.

11.1.5 Typical Technique

A sample is introduced, either manually or with an autosampler, into a sample loop of known volume. A buffered aqueous solution known as the mobile phase carries the sample from the loop onto a column that contains some form of stationary phase material. This is typically a resin or gel matrix consisting of agarose or cellulose beads with covalently bonded charged functional groups. Equilibration of the stationary phase is needed in order to obtain the desired charge of the column. If the column is not properly equilibrated the desired molecule may not bind strongly to the column. The target analytes (anions or cations) are retained on the stationary phase but can be eluted by increasing the concentration of a similarly charged species that displaces the analyte ions from the stationary phase. For example, in cation exchange chromatography, the positively charged analyte can be displaced by adding positively charged sodium ions. The analytes of interest must then be detected by some means, typically by conductivity or UV/visible light absorbance. Control an IC system usually requires a chromatography data system (CDS). In addition to IC systems, some of these CDSs can also control gas chromatography (GC) and HPLC.

11.1.6 Membrane Exchange Chromatography

A type of ion exchange chromatography, membrane exchange is a relatively new method of purification designed to overcome limitations of using columns packed with beads. Membrane Chromatographic devices are cheap to mass-produce and disposable unlike other chromatography devices that require maintenance and time to revalidate. There are three types of membrane absorbers that are typically used when separating substances. The three types are flat sheet, hollow fibre, and radial flow. The most common absorber and best suited for membrane chromatography is multiple flat sheets because it has more absorbent volume. It can be used to overcome mass transfer limitations and pressure drop, making it especially advantageous for isolating and purifying viruses, plasmid DNA, and other large macromolecules. The column is packed with microporous membranes with internal pores which contain adsorptive moieties that can bind the target protein. Adsorptive membranes are available in a variety of geometries and chemistry which allows them to be used for purification and also fractionation, concentration, and clarification in an efficiency that is 10 fold that of using beads. Membranes can be prepared through isolation of the membrane itself, where membranes are cut into squares and immobilized. A more recent method involved the use of live cells that are attached to a support membrane and are used for identification and clarification of signaling molecules.

11.1.7 Separating Proteins

Ion exchange chromatography can be used to separate proteins because they contain charged functional groups. The ions of interest (in this case charged proteins) are exchanged for another ions (usually H^+) on a charged solid support. The solutes are most commonly in a liquid phase, which tends to be water. Take for example proteins in water, which would be a liquid phase that is passed through a column. The column is commonly known as the solid phase since it is filled with porous synthetic particles that are of a particular charge. These porous particles are also referred to as beads, may be aminated (containing amino groups) or have metal ions in order to have a charge. The column can be prepared using porous polymers, for macromolecules of a mass of over 100 000 Da, the optimum size of the porous particle is about 1 μm^2. This is because slow diffusion of the solutes within the pores does not restrict the separation quality. The beads containing positively charged groups, which attract the negatively charged proteins, are commonly referred to as anion exchange resins. The amino acids that have negatively charged side chains at pH 7 (pH of water) are glutamate and aspartate. The beads that are negatively charged are called cation exchange resins, as positively charged proteins will be attracted. The amino acids that have positively charged side chains at pH 7 are lysine, histidine and arginine.

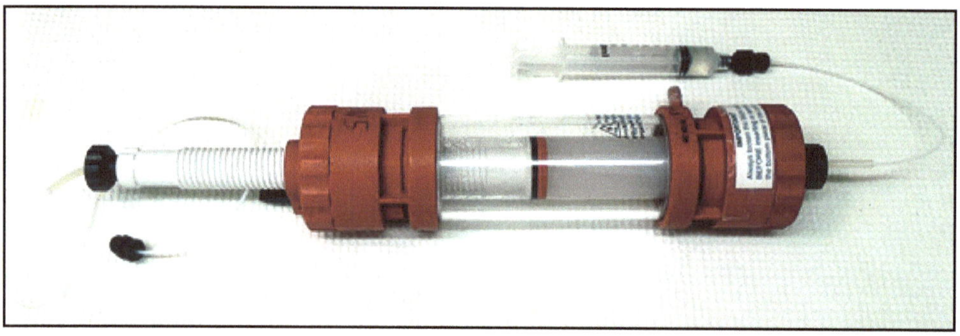

Fig: Preparative-Scale Ion Exchange Column Used for Protein Purification

The isoelectric point is the pH at which a compound - in this case a protein - has no net charge. A protein's isoelectric point or PI can be determined using the pKa of the side chains, if the amino (positive chain) is able to cancel out the carboxyl (negative) chain, the protein would be at its PI. Using buffers instead of water for proteins that do not have a charge at pH 7 is a good idea as it enables the manipulation of pH to alter ionic interactions between the proteins and the beads. Weakly acidic or basic side chains are able to have a charge if the pH is high or low enough respectively. Separation can be achieved based on the natural isoelectric point of the protein. Alternatively a peptide tag can be genetically added to the protein to give the protein an isoelectric point away from most natural proteins (e.g., 6 arginines for binding to a cation-exchange resin or 6 glutamates for binding to an anion-exchange resin such as DEAE-Sepharose).

Elution by increasing ionic strength of the mobile phase is more subtle. It works because ions from the mobile phase interact with the immobilized ions on the stationary phase, thus "shielding" the stationary phase from the protein, and letting the protein elute. Elution from ion-exchange columns can be sensitive to changes of a single charge- chromatofocusing. Ion-exchange chromatography is also useful in the isolation of specific multimeric protein assemblies, allowing purification of specific complexes according to both the number and the position of charged peptide tags.

Gibbs–Donnan Effect

In ion exchange chromatography, the Gibbs–Donnan effect is observed when the pH of the applied buffer and the ion exchanger differ, even up to one pH unit. For example, in anion-exchange columns, the ion exchangers repeal protons so the pH of the buffer near the column differs is higher than the rest of the solvent. As a result, an experimenter has to be careful that the protein(s) of interest is stable and properly charged in the "actual" pH. This effect comes as a result of two similarly charged particles, one from the resin and one from the solution, failing to distribute properly between the two sides; there is a selective uptake of one ion over another. For example, in a sulphonated polystyrene resin, a cation exchange resin, the chlorine ion of a hydrochloric acid buffer should equilibrate into the resin. However, since the concentration of the sulphonic acid in the resin is high, the hydrogen of HCl has no tendency to enter the column. This, combined with the need of electroneutrality, leads to a minimum amount of hydrogen and chlorine entering the resin.

11.1.8 Uses and Practical Applications of Ion Exchange Chromatography

Clinical Utility

A use of ion chromatography can be seen in argentation chromatography. Usually, silver and compounds containing acetylenic and ethylenic bonds have very weak interactions. This phenomenon has been widely tested on olefin compounds. The ion complexes the olefins make with silver ions are weak and made based on the overlapping of pi, sigma, and d orbitals and available electrons therefore cause no real changes in the double bond. This behavior was manipulated to separate lipids, mainly fatty acids from mixtures in to fractions with differing number of double bonds using silver ions. The ion resins were impregnated with silver ions, which were then exposed to various acids (silicic acid) to elute fatty acids of different characteristics.

Detection limits as low as 1 μM can be obtained for alkali metal ions. It may be used for measurement of HbA1c, porphyrin and with water purification. Ion Exchange Resins(IER) have been widely used especially in medicines due to its high capacity and the uncomplicated system of the separation process. One of the synthetic uses is to use Ion Exchange Resins for kidney dialysis. This method is used to separate the blood elements by using the cellulose membraned artificial kidney.

Another clinical application of ion chromatography is in the rapid anion exchange chromatography technique used to separate creatine kinase (CK) isoenzymes from human serum and tissue sourced in autopsy material (mostly CK rich tissues were used such as cardiac muscle and brain).

These isoenzymes include MM, MB, and BB, which all carry out the same function given different amino acid sequences. The functions of these isoenzymes are to convert creatine, using ATP, into phosphocreatine expelling ADP. Mini columns were filled with DEAE-Sephadex A-50 and further eluted with tris- buffer sodium chloride at various concentrations (each concentration was chosen advantageously to manipulate elution).

Human tissue extract was inserted in columns for separation. All fractions were analyzed to see total CK activity and it was found that each source of CK isoenzymes had characteristic isoenzymes found within. Firstly, CK- MM was eluted, then CK-MB, followed by CK-BB. Therefore, the isoenzymes found in each sample could be used to identify the source, as they were tissue specific.

Using the information from results, correlation could be made about the diagnosis of patients and the kind of CK isoenzymes found in most abundant activity. From the finding, about 35 out of 71 patients studied suffered from heart attack (myocardial infarction) also contained an abundant amount of the CK-MM and CK-MB isoenzymes. Findings further show that many other diagnosis including renal failure, cerebrovascular disease, and pulmonary disease were only found to have the CK-MM isoenzyme and no other isoenzyme. The results from this study indicate correlations between various diseases and the CK isoenzymes found which confirms previous test results using various techniques. Studies about CK-MB found in heart attack victims have expanded since this study and application of ion chromatography.

Industrial Applications

Since 1975 ion chromatography has been widely used in many branches of industry. The main beneficial advantages are reliability, very good accuracy and precision, high selectivity, high speed, high separation efficiency, and low cost of consumables. The most significant development related to ion chromatography are new sample preparation methods; improving the speed and selectivity of analytes separation; lowering of limits of detection and limits of quantification; extending the scope of applications; development of new standard methods; miniaturization and extending the scope of the analysis of a new group of substances. Allows for quantitative testing of electrolyte and proprietary additives of electroplating baths.

It is an advancement of qualitative hull cell testing or less accurate UV testing. Ions, catalysts, brighteners and accelerators can be measured. Ion exchange chromatography has gradually become a widely known, universal technique for the detection of both anionic and cationic species. Applications for such purposes have been developed, or are under development, for a variety of fields of interest, and in particular, the pharmaceutical industry. The usage of ion exchange chromatography in pharmaceuticals has increased in recent years, and in 2006, a chapter on ion exchange chromatography was officially added to the United States Pharmacopia-National Formulary (USP-NF). Furthermore, in 2009 release of the USP-NF, the United States Pharmacopia made several analyses of ion chromatography available using two techniques: conductivity detection, as well as pulse amperometric detection. Majority of these applications are primarily used for measuring and analyzing residual limits in pharmaceuticals, including detecting the limits of oxalate, iodide, sulfate, sulfamate, phosphate, as well as various electrolytes including potassium, and sodium. In total, the 2009 edition of the USP-NF officially released twenty eight methods of detection for the analysis of active compounds, or components of active compounds, using either conductivity detection or pulse amperometric detection.

Drug Development

There has been a growing interest in the application of IC in the analysis of pharmaceutical drugs. IC is used in different aspects of product development and quality control testing. For example, IC is used to improve stabilities and solubility properties of pharmaceutical active drugs molecules as well as used to detect systems that have higher tolerance for organic solvents. IC has been used for the determination of analytes as a part of a dissolution test. For instance, calcium dissolution tests have shown that other ions present in the medium can be well resolved among themselves and also from the calcium ion. Therefore, IC has been employed in drugs in the form of tablets and capsules in order to determine the amount of drug dissolve with time. IC is also widely used for detection and quantification of excipients or inactive ingredients used in pharmaceutical formulations. Detection of sugar and sugar alcohol in such formulations through IC has been done due to these polar groups getting resolved in ion column. IC methodology also established in analysis of impurities in drug substances and products. Impurities or any components that are not part of the drug chemical entity are evaluated and they give insights about the maximum and minimum amounts of drug that should be administered in a patient per day.

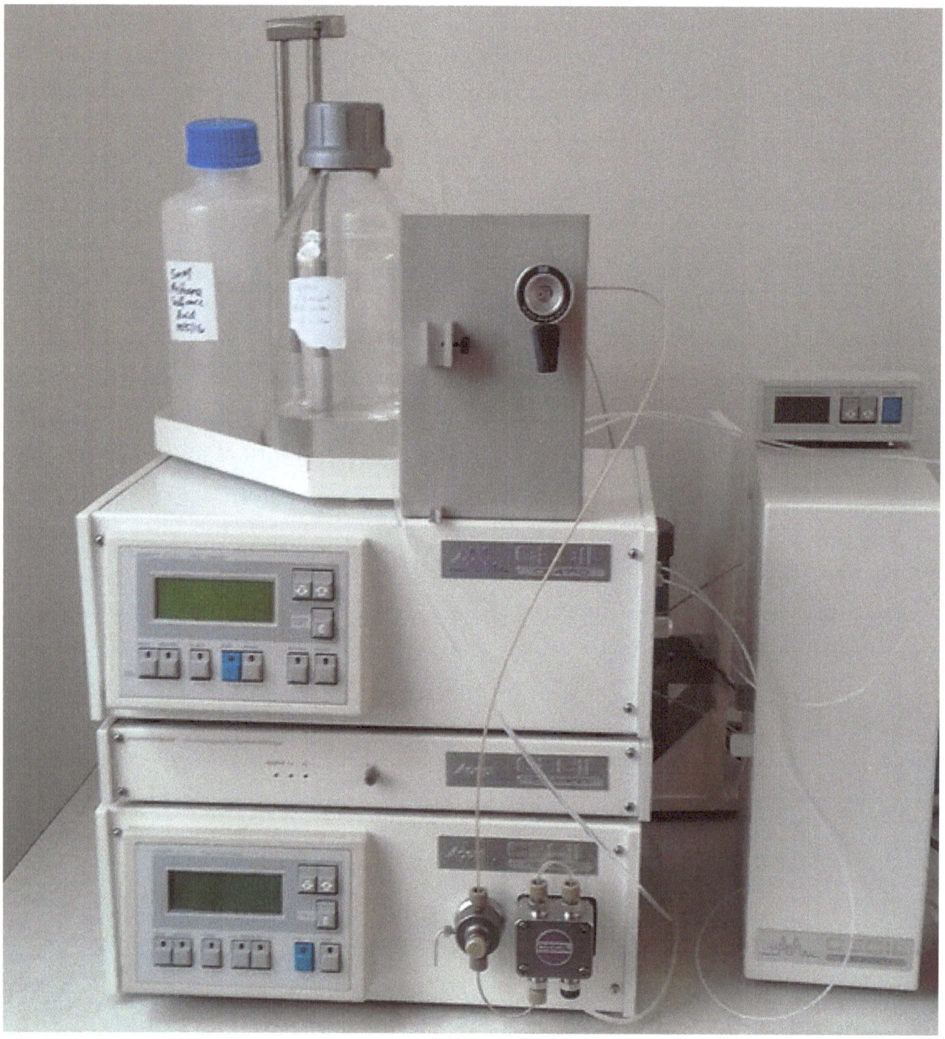

Fig: An Ion Chromatography System Used to Detect and Measure Cations such as Sodium, Ammonium and Potassium in Expectorant Cough Formulations

11.1.9 Advantages of Ion Exchange Chromatography

One of the major strengths of ion exchange chromatography lies in its high binding capacity, which enables the purification of large quantities of biomolecules in a relatively small volume of resin. This makes the technique well-suited for both analytical and preparative scales, from microgram to kilogram quantities.

The method also provides excellent resolution, allowing the separation of closely related molecules based on subtle differences in charge. It is highly reproducible, with well-defined operating parameters that can be easily optimized and scaled up. The use of aqueous buffers makes it compatible with biologically active molecules, preserving their structure and functionality during the purification process.

Another advantage is the ability to integrate ion exchange steps seamlessly into multi-modal purification protocols. It can be used in combination with other chromatographic techniques such as gel filtration or affinity chromatography to achieve greater purity and specificity. Furthermore, modern ion exchange columns are robust, chemically stable, and suitable for multiple cycles of regeneration and reuse, enhancing cost-effectiveness.

11.1.10 Challenges and Considerations

Despite its many benefits, ion exchange chromatography has some limitations that must be carefully managed. One challenge is the sensitivity of the separation to changes in buffer pH and ionic strength. Small variations in these parameters can significantly affect the binding behavior and elution profile, necessitating precise control and standardization.

Another limitation is the potential for non-specific interactions, especially with hydrophobic or highly charged proteins, which can lead to peak broadening or reduced resolution. In such cases, additives such as detergents, organic solvents, or specific ions may be required to minimize undesired effects.

The method also requires prior knowledge of the isoelectric point and stability characteristics of the target molecule in order to design an effective purification strategy. In some complex mixtures, overlapping peaks or co-elution of similarly charged molecules can occur, making separation difficult without additional purification steps.

11.1.11 Recent Developments and Future Prospects

Recent advances in ion exchange chromatography have focused on improving resin performance, enhancing resolution, and increasing throughput. New-generation resins feature improved particle uniformity, higher porosity, and reduced backpressure, enabling faster separations and better column efficiency. High-performance IEX columns with smaller particle sizes have been developed for analytical applications such as charge variant analysis in monoclonal antibodies and biosimilars. Automated chromatography systems with sophisticated gradient control and real-time monitoring have further enhanced the reproducibility and scalability of IEX processes. In addition, the integration of ion exchange chromatography with advanced detection methods such as mass spectrometry and multi-angle light scattering is opening new avenues for detailed biomolecular characterization.

In the future, the development of novel ion exchange media with tailored selectivity, improved chemical stability, and higher dynamic binding capacities is expected to broaden the applicability of this technique. As the demand for high-purity biotherapeutics and precision biomolecular analysis continues to grow, ion exchange chromatography will remain a vital tool in the bioseparation arsenal.

11.2 CHROMATOFOCUSING

Chromatofocusing is a protein-separation technique that allows resolution of single proteins and other ampholytes from a complex mixture according to differences in their isoelectric point. Chromatofocusing uses ion exchange resins and is typically performed on fast protein liquid chromatography (FPLC) or similar equipment capable of producing continuous buffer gradients, though this is not a requirement. In contrast to typical ion exchange chromatography, where bound molecules are eluted from the resin by increasing the ionic strength of the buffer environment, chromatofocusing elutes bound species by altering the pH of the buffer. This changes the net surface charge of bound molecules, altering their avidity for the resin. As the changing pH of the buffer system traverses the pI of a given molecule, that molecule will elute from the resin as it will no longer possess a net surface charge (a requisite for molecular binding to ion exchange resins). Chromatofocusing is a powerful purification technique with respect to proteins as it can resolve very similar species differing by less than 0.05 pH units that may not separate well, or at all, using traditional ion exchange strategies. A major drawback to this technique is that some proteins will aggregate when they are present at relatively high concentrations and carry no net surface charge. This can cause blockage of the resin, which is highly problematic when using sealed columns of ion exchange resin on FPLC equipment, resulting in pressure buildup and possible equipment failure. Apparent aggregation issues can sometimes be overcome by limiting the sample concentration and use of buffer additives that deter aggregate formation.

11.2.1 Principle and Mechanism of Chromatofocusing

The basic principle of chromatofocusing involves the use of an ion exchange column that is pre-equilibrated with a buffer of relatively high pH. Proteins are then loaded onto this column and allowed to bind via electrostatic interactions. Following this, a buffering solution with a lower pH is applied to the column. As the elution progresses, a pH gradient is gradually formed along the column length.

Proteins that are bound to the ion exchange resin will elute at the point where the pH of the gradient matches their isoelectric point, the specific pH at which they carry no net electrical charge and thus lose their affinity for the charged resin.

At the molecular level, chromatofocusing exploits the amphoteric nature of proteins, which can accept or donate protons depending on the surrounding pH. As the pH gradient migrates through the column, proteins lose their net positive or negative charges at their isoelectric points and are displaced from the resin. Because this transition is very sensitive to small changes in pH, the technique provides excellent resolution between closely related species.

11.2.2 Column Materials and Buffer Systems

Chromatofocusing typically utilizes anion exchange resins as the stationary phase, particularly those with weakly basic functional groups such as diethylaminoethyl (DEAE) or quaternary amines. These groups interact strongly with negatively charged proteins when the column is initially equilibrated with an alkaline buffer. The choice of resin must ensure minimal non-specific binding and stability across a wide pH range, usually from pH 9 to 4, to accommodate the full spectrum of protein isoelectric points.

The success of chromatofocusing depends heavily on the selection of buffer systems capable of forming a smooth and stable pH gradient. Commonly used buffer components include polyampholytes, which are mixtures of amphoteric compounds with buffering capacities spanning a broad pH range. These polyampholytes are particularly effective in establishing a continuous and reproducible gradient without excessive conductivity, making them ideal for this technique. Alternatively, pH gradients can be generated by titrating one buffer with another or by using commercially available pH gradient kits specifically designed for chromatofocusing.

11.2.3 Applications in Protein Separation and Purification

Chromatofocusing is particularly valuable for separating proteins that are difficult to resolve by other chromatographic methods, such as those differing only slightly in charge or structure. It is widely used for the purification and analysis of protein isoforms, monoclonal antibodies, enzymes, and post-translationally modified variants. These forms often have nearly identical molecular weights but differ slightly in their net surface charge due to variations in amino acid sequence, glycosylation, phosphorylation, or other modifications.

In pharmaceutical biotechnology, chromatofocusing is frequently employed to characterize charge heterogeneity in monoclonal antibody preparations. Since small changes in the pI of antibodies can indicate modifications that affect their therapeutic activity or stability, chromatofocusing is a valuable quality control tool. It helps in the detection of deamidation, sialylation, and other biochemical modifications that are critical in regulatory assessment and product development.

Chromatofocusing is also beneficial in basic research where understanding protein isoform expression or identifying biomarkers is essential. For example, in proteomics studies, chromatofocusing can be used as a preparative step before mass spectrometry to reduce sample complexity and enhance detection sensitivity.

11.2.4 Advantages of Chromatofocusing

One of the primary advantages of chromatofocusing is its ability to separate proteins with high resolution based solely on their isoelectric points. This makes it uniquely suited for applications where small differences in pI need to be distinguished, such as separating closely related isoforms or detecting slight modifications. Because elution occurs at the precise isoelectric point of a protein, the method provides information that can be directly correlated with protein identity and purity.

Another advantage is the gentle, non-denaturing nature of the technique. Since the separation relies on pH changes rather than salt gradients or organic solvents, proteins generally retain their native conformation and activity. This is particularly important for enzymes, antibodies, and other bioactive compounds that are sensitive to harsh conditions.

The use of low-conductivity elution buffers also simplifies downstream processing, especially when preparing samples for electrophoresis, crystallography, or mass spectrometry. This reduces the need for desalting or buffer exchange steps, thereby streamlining the purification workflow.

11.2.5 Challenges and Limitations

Despite its strengths, chromatofocusing presents certain challenges that must be addressed to ensure optimal performance. One of the main difficulties lies in maintaining a stable and linear pH gradient throughout the column. Inconsistent gradients can lead to poor resolution, overlapping peaks, or incomplete separation. The gradient formation is influenced by the flow rate, column packing, and buffering capacity of the polyampholyte mixture, all of which require careful optimization.

Another limitation is the relatively low loading capacity compared to conventional ion exchange chromatography. Because the separation depends on narrow differences in pI, overloading the column can lead to peak broadening and diminished resolution. This restricts the use of chromatofocusing in large-scale purification processes unless coupled with other high-capacity techniques.

Additionally, the method is highly sensitive to the pH stability of proteins. If the target biomolecules are prone to denaturation or precipitation within the pH range of the gradient, they may lose activity or be lost during elution.

Therefore, knowledge of the protein's stability profile is crucial when designing a chromatofocusing experiment.

Instrumentation and monitoring can also be a limiting factor. Since elution is based on pH rather than absorbance or conductivity alone, precise control and real-time monitoring of the gradient are essential. Modern chromatography systems with integrated pH sensors and automated gradient control can help mitigate these issues, but they may not always be available in basic laboratories.

11.2.6 Recent Developments and Technological Enhancements

Recent innovations in chromatofocusing technology aim to overcome its traditional limitations and expand its application scope. High-performance chromatofocusing columns with optimized packing and improved flow characteristics have been introduced, offering better gradient stability and higher resolution. These columns are also compatible with fast protein liquid chromatography (FPLC) systems, enabling automation and real-time monitoring.

The development of synthetic polyampholytes and proprietary pH gradient kits has further improved the reproducibility and ease of use of chromatofocusing. These formulations allow for fine-tuned gradient control and consistent performance across multiple runs, which is particularly valuable in regulated environments such as pharmaceutical manufacturing.

Advances in tandem purification strategies, where chromatofocusing is combined with other techniques like reverse-phase or affinity chromatography, are enabling more comprehensive and efficient purification schemes. These integrated approaches enhance selectivity and throughput, particularly in complex biological samples.

CHAPTER 12

Reversed Phase and Hydrophobic Interaction Chromatography

Chromatographic techniques form the backbone of modern bioseparations, offering high resolution, selectivity, and efficiency in the purification of biomolecules. Among these, Reversed Phase Chromatography (RPC) and Hydrophobic Interaction Chromatography (HIC) stand out as two prominent methods that exploit the hydrophobic nature of biological molecules for their separation. Though both techniques are based on hydrophobic interactions, they differ in their operating principles, mobile and stationary phases, and the nature of biological samples they are best suited to analyze. Together, RPC and HIC provide a versatile toolkit for researchers and industries dealing with proteins, peptides, nucleotides, and small biomolecules, offering essential support for applications ranging from drug development to proteomics and quality control.

12.1 FUNDAMENTALS OF HYDROPHOBIC INTERACTIONS

The core concept that underlies both RPC and HIC is the hydrophobic interaction, a non-covalent force that drives the association of non-polar regions of molecules in aqueous environments. These interactions are entropically favorable, as they minimize the disruption of the hydrogen-bonding network of water. In chromatography, these forces are utilized to temporarily retain molecules on hydrophobic surfaces, allowing for their differential elution based on the strength and extent of hydrophobicity. However, the way these interactions are modulated—either through salt concentration in HIC or organic solvent concentration in RPC—differentiates the two methods significantly.

12.2 HYDROPHOBIC INTERACTION CHROMATOGRAPHY (HIC): PRINCIPLE AND OPERATION

Hydrophilic interaction chromatography (or hydrophilic interaction liquid chromatography, HILIC) is a variant of normal phase liquid chromatography that partly overlaps with other chromatographic applications such as ion chromatography and reversed phase liquid chromatography. HILIC uses hydrophilic stationary phases with reversed-phase type eluents. The name was suggested by Andrew Alpert in his 1990 paper on the subject. He described the chromatographic mechanism for it as liquid-liquid partition chromatography where analytes elute in order of increasing polarity, a conclusion supported by a review and re-evaluation of published data.

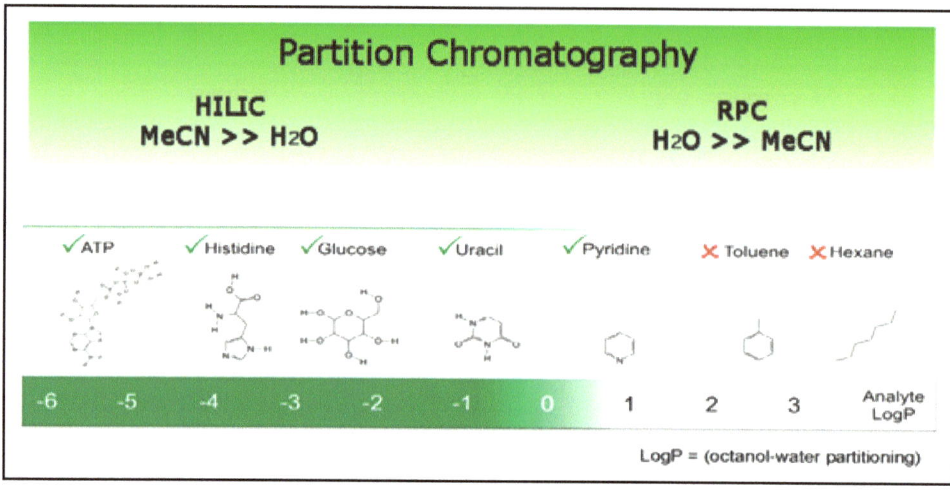

Fig: HILIC Partition Technique Useful Range

12.2.1 Surface

Any polar chromatographic surface can be used for HILIC separations. Even non-polar bonded silicas have been used with extremely high organic solvent composition, thanks to the exposed patches of silica in between the bonded ligands on the support, which can affect the interactions. With that exception, HILIC phases can be grouped into five categories of neutral polar or ionic surfaces:

- Simple unbonded silica silanol or diol bonded phases.
- Amino or anionic bonded phases.
- Amide bonded phases.
- Cationic bonded phases.
- Zwitterionic bonded phases.

12.2.2 Mobile Phase

A typical mobile phase for HILIC chromatography includes acetonitrile ("MeCN", also designated as "ACN") with a small amount of water. However, any aprotic solvent miscible with water (e.g. THF or dioxane) can be used. Alcohols can also be used, however, their concentration must be higher to achieve the same degree of retention for an analyte relative to an aprotic solvent–water combination.

It is commonly believed that in HILIC, the mobile phase forms a water-rich layer on the surface of the polar stationary phase vs. the water-deficient mobile phase, creating a liquid/liquid extraction system. The analyte is distributed between these two layers. However, HILIC is more than just simple partitioning and includes hydrogen donor interactions between neutral polar species as well as weak electrostatic mechanisms under the high organic solvent conditions used for retention. This distinguishes HILIC as a mechanism distinct from ion exchange chromatography. The more polar compounds will have a stronger interaction with the stationary aqueous layer than the less polar compounds. Thus, a separation based on a compound's polarity and degree of solvation takes place.

12.2.3 Additives

Ionic additives, such as ammonium acetate and ammonium formate, are usually used to control the mobile phase pH and ion strength. In HILIC they can also contribute to the polarity of the analyte, resulting in differential changes in retention. For extremely polar analytes (e.g. aminoglycoside antibiotics (gentamicin) or adenosine triphosphate), higher concentrations of buffer (c. 100 mM) are required to ensure that the analyte will be in a single ionic form. Otherwise, asymmetric peak shape, chromatographic tailing, and/or poor recovery from the stationary phase will be observed. For the separation of neutral polar analytes (e.g. carbohydrates), no buffer is necessary.

Other salts, such as 100–300 mM sodium perchlorate, that are soluble in high-organic solvent mixtures (c. 70–90% acetonitrile), can be used to increase the mobile phase polarity to affect elution These salts are not volatile, so this technique is less useful with a mass spectrometer as the detector. Usually a gradient (to increasing amounts of water) is enough to promote elution.

All ions partition into the stationary phase to some degree, so an occasional "wash" with water is required to ensure a reproducible stationary phase.

12.2.4 Applications

The HILIC mode of separation is used extensively for separation of some biomolecules, organic and some inorganic molecules by differences in polarity. Its utility has increased due to the simplified sample preparation for biological samples, when analyzing for metabolites, since the metabolic process generally results in the addition of polar groups to enhance elimination from the cellular tissue.

This separation technique is also particularly suitable for glycosylation analysis and quality assurance of glycoproteins and glycoforms in biologic medical products. For the detection of polar compounds with the use of electrospray-ionization mass spectrometry as a chromatographic detector, HILIC can offer a ten fold increase in sensitivity over reversed-phase chromatography because the organic solvent is much more volatile.

12.2.5 Choice of Ph

With surface chemistries that are weakly ionic, the choice of pH can affect the ionic nature of the column chemistry. Properly adjusted, the pH can be set to reduce the selectivity toward functional groups with the same charge as the column, or enhance it for oppositely charged functional groups. Similarly, the choice of pH affects the polarity of the solutes. However, for column surface chemistries that are strongly ionic, and thus resistant to pH values in the mid-range of the pH scale (pH 3.5–8.5), these separations will be reflective of the polarity of the analytes alone, and thus might be easier to understand when doing methods development.

12.2.6 ERLIC

In 2008, Alpert coined the term, ERLIC (electrostatic repulsion hydrophilic interaction chromatography), for HILIC separations where an ionic column surface chemistry is used to repel a common ionic polar group on an analyte or within a set of analytes, to facilitate separation by the remaining polar groups. Electrostatic effects have an order of magnitude stronger chemical potential than neutral polar effects. This allows one to minimize the influence of a common, ionic group within a set of analyte molecules; or to reduce the degree of retention from these more polar functional groups, even enabling isocratic separations in lieu of a gradient in some situations. His subsequent publication further described orientation effects which others have also called ion-pair normal phase or e-HILIC, reflecting retention mechanisms sensitive to a particular ionic portion of the analyte, either attractive or repulsive. ERLIC (eHILIC) separations need not be isocratic, but the net effect is the reduction of the attraction of a particularly strong polar group, which then requires less strong elution conditions, and the enhanced interaction of the remaining polar (opposite charged ionic, or non-ionic) functional groups of the analyte(s).Based on the ERLIC column invented by Andrew Alpert, a new peptide mapping methodology was developed with unique properties of separation of asparagine deamidation and isomerization. This unique properties would be very beneficial for future mass spectrometry based multi-attributes monitoring in biologics quality control.

12.2.7 Cationic eHILIC

For example, one could use a cation exchange (negatively charged) surface chemistry for ERLIC separations to reduce the influence on retention of anionic (negatively charged) groups (the phosphates of nucleotides or of phosphonyl antibiotic mixtures; or sialic acid groups of modified carbohydrates) to now allow separation based more on the basic and/or neutral functional groups of these molecules. Modifying the polarity of a weakly ionic group (e.g. carboxyl) on the surface is easily accomplished by adjusting the pH to be within two pH units of that group's pKa. For strongly ionic functional groups of the surface (i.e. sulfates or phosphates) one could instead use a lower amount of buffer so the residual charge is not completely ion paired. An example of this would be the use of a 12.5mM (rather than the recommended >20mM buffer), pH 9.2 mobile phase on a polymeric, zwitterionic, betaine-sulfonate surface to separate phosphonyl antibiotic mixtures (each containing a phosphate group). This enhances the influence of the column's sulfonic acid functional groups of its surface chemistry over its, slightly diminished (by pH), quaternary amine. Commensurate with this, these analytes will show a reduced retention on the column eluting earlier, and in higher amounts of organic solvent, than if a neutral polar HILIC surface were used. This also increases their detection sensitivity by negative ion mass spectrometry.

12.2.8 Anionic eHILIC

One can use an anion exchange (positively charged) column surface chemistry to reduce the influence on retention of cationic (positively charged) functional groups for a set of analytes, such as when selectively isolating phosphorylated peptides or sulfated polysaccharide molecules. Use of a pH between 1 and 2 pH units will reduce the polarity of two of the three ionizable oxygens of the phosphate group, and thus will allow easy desorption from the (oppositely charged) surface chemistry. It will also reduce the influence of negatively charged carboxyls in the analytes, since they will be protonated at this low a pH value, and thus contribute less overall polarity to the molecule. Any common, positively charged amino groups will be repelled from the column surface chemistry and thus these conditions enhance the role of the phosphate's polarity (as well as other neutral polar groups) in the separation.

12.2.9 Applications and Advantages of HIC

Hydrophobic Interaction Chromatography is widely used in the downstream processing of biopharmaceuticals, especially monoclonal antibodies, recombinant proteins, and enzymes. It is often employed in intermediate purification steps following precipitation or affinity chromatography. One of its main advantages is the ability to operate under physiological pH and ionic strength, which helps maintain protein stability. Furthermore, HIC can serve as an orthogonal method to ion exchange or affinity chromatography, adding another layer of selectivity based on hydrophobic properties.

HIC is also valuable in the characterization of protein variants, aggregates, and degradation products. By analyzing the retention time and peak patterns, researchers can gain insights into conformational changes or the presence of hydrophobic modifications. These insights are critical in quality control and formulation development in the pharmaceutical industry.

12.2.10 Limitations of HIC

Despite its advantages, HIC does have limitations. One major challenge is the reliance on high concentrations of salt at the beginning of the process, which can complicate downstream desalting or interfere with certain analytical techniques. The resolution of closely related hydrophobic species can also be limited compared to more sensitive methods such as reversed phase chromatography. In addition, the selection of the appropriate salt and stationary phase requires careful optimization, as different proteins may respond differently to similar conditions.

12.3 REVERSED PHASE CHROMATOGRAPHY (RPC): PRINCIPLE AND MECHANISM

Reversed-phase liquid chromatography (RP-LC) is a mode of liquid chromatography in which non-polar stationary phase and polar mobile phases are used for the separation of organic compounds. The vast majority of separations and analyses using high-performance liquid chromatography (HPLC) in recent years are done using the reversed phase mode. In the reversed phase mode, the sample components are retained in the system the more hydrophobic they are.

The factors affecting the retention and separation of solutes in the reversed phase chromatographic system are as follows:

- The chemical nature of the stationary phase, i.e., the ligands bonded on its surface, as well as their bonding density, namely the extent of their coverage.
- The composition of the mobile phase. Type of the bulk solvents whose mixtures affect the polarity of the mobile phase, hence the name modifier for a solvent added to affect the polarity of the mobile phase.
- Additives, such as buffers, affect the pH of the mobile phase, which affect the ionization state of the solutes and their polarity.

In order to retain the organic components in mixtures, the stationary phases, packed within columns, consist of a hydrophobic substrates, bonded to the surface of porous silica-gel particles in various geometries (spheric, irregular), at different diameters (sub-2, 3, 5, 7, 10 um), with varying pore diameters (60, 100, 150, 300, A). The particle's surface is covered by chemically bonded hydrocarbons, such as C3, C4, C8, C18 and more. The longer the hydrocarbon associated with the stationary phase, the longer the sample components will be retained. Some stationary phases are also made of hydrophobic polymeric particles, or hybridized silica-organic groups particles, for method in which mobile phases at extreme pH are used.

Most current methods of separation of biomedical materials use C-18 columns, sometimes called by trade names, such as ODS (octadecylsilane) or RP-18.

The mobile phases are mixtures of water and polar organic solvents, the vast majority of which are methanol and acetonitrile. These mixtures usually contain various additives such as buffers (acetate, phosphate, citrate), surfactants (alkyl amines or alkyl sulfonates) and special additives (EDTA). The goal of using supplements of one kind or another is to increase efficiency, selectivity, and control solute retention.

12.3.1 Stationary Phases

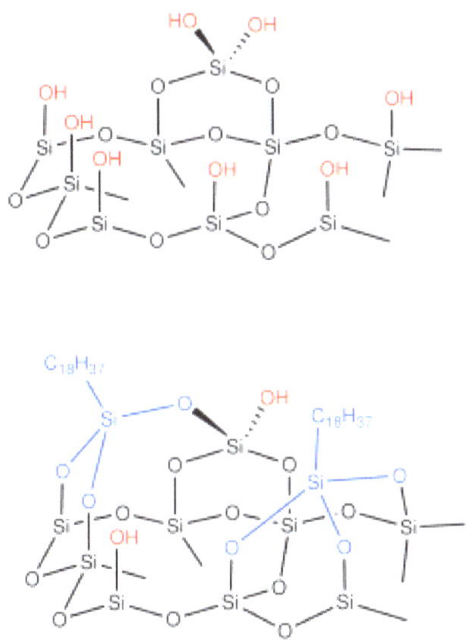

Fig: Idealized Cartoon of Silica Gel before and after Treatment with Octadecyltrichlorosilane ($C_{18}H_{37}SiCl_3$. Most Silanol Groups (Red) are Converted to Hydrophobic Alkylsiloxy Groups

In the 1970s, most liquid chromatography runs were performed using solid particles as the stationary phases, made of unmodified silica gel or alumina. This type of technique is now referred to as normal-phase chromatography. Since the stationary phase is hydrophilic in this technique, and the mobile phase is non-polar (consisting of organic solvents such as hexane and heptane), biomolecules with hydrophilic properties in the sample adsorb to the stationary phase strongly. Moreover, they were not dissolved easily in the mobile phase solvents. At the same time hydrophobic molecules experience less affinity to the polar stationary phase, and elute through it early with not enough retention. This was the reasons why during the 1970s the silica based particles were treated with hydrocarbons, immobilized or bonded on their surface, and the mobile phases were switched to aqueous and polar in nature, to accommodate biomedical substances.

The use of a hydrophobic stationary phase and polar mobile phases is essentially the reverse of normal phase chromatography, since the polarity of the mobile and stationary phases have been inverted – hence the term reversed-phase chromatography. As a result, hydrophobic molecules in the polar mobile phase tend to adsorb to the hydrophobic stationary phase, and hydrophilic molecules in the sample pass through the column and are eluted first. Hydrophobic molecules can be eluted from the column by decreasing the polarity of the mobile phase using an organic (non-polar) solvent, which reduces hydrophobic interactions. The more hydrophobic the molecule, the more strongly it will bind to the stationary phase, and the higher the concentration of organic solvent that will be required to elute the molecule.

Many of the mathematical parameters of the theory of chromatography and experimental considerations used in other chromatographic methods apply to RP-LC as well (for example, the selectivity factor, chromatographic resolution, plate count, etc. It can be used for the separation of a wide variety of molecules. It is typically used for separation of proteins, because the organic solvents used in normal-phase chromatography can denature many proteins.

Today, RP-LC is a frequently used analytical technique. There are huge variety of stationary phases available for use in RP-LC, allowing great flexibility in the development of the separation methods.

Silica-Based Stationary Phases

Silica gel particles are commonly used as a stationary phase in high-performance liquid chromatography (HPLC) for several reasons, including:

- **High Surface Area:** Silica gel particles have a high surface area, allowing direct interactions with solutes or after bonding of variety of ligands for versatile interactions with the sample molecules, leading to better separations.
- **Chemical and Thermal Stability and Inertness:** Silica gel is chemically stable, as it usually does not react with either the solvents of the mobile phase nor the compounds being separated, resulting in accurate, repeatable and reliable analyses.
- **Wide Applicability:** Silica gel is versatile and can be modified with various functional groups, making it suitable for a wide range of analytes and applications.
- **Efficient Separation:** The unique properties of silica gel particles, combined with their high surface area and controlled average particle diameter pore size, facilitate efficient and precise separation of compounds in HPLC.

- **Reproducibility:** Silica gel particles can offer high batch-to-batch reproducibility, which is crucial for consistent and reliable HPLC analyses throughout decades.
- **Particle Diameter and Pore Size Control:** Silica gel can be engineered to have specific pore sizes, enabling precise control over separation based on molecular size.
- **Cost-Effectiveness:** Silica is the most abundant element on earth, hence its gel is a cost-effective choice for HPLC applications, making it widely adopted in laboratories.

The United States Pharmacopoeia (USP) has classified HPLC columns by L# types. The most popular column in this classification is an octadecyl carbon chain (C18)-bonded silica (USP classification L1). This is followed by C8-bonded silica (L7), pure silica (L3), cyano-bonded silica (CN) (L10) and phenyl-bonded silica (L11). Note that C18, C8 and phenyl are dedicated reversed-phase stationary phases, while CN columns can be used in a reversed-phase mode depending on analyte and mobile phase conditions. Not all C18 columns have identical retention properties. Surface functionalization of silica can be performed in a monomeric or a polymeric reaction with different short-chain organosilanes used in a second step to cover remaining silanol groups (end-capping). While the overall retention mechanism remains the same, subtle differences in the surface chemistries of different stationary phases will lead to changes in selectivity.

Modern columns have different polarity depending on the ligand bonded to the stationary phase. PFP is pentafluorphenyl. CN is cyano. NH2 is amino. ODS is octadecyl or C18. ODCN is a mixed mode column consisting of C18 and nitrile.

Recent developments in chromatographic supports and instrumentation for liquid chromatography (LC) facilitate rapid and highly efficient separations, using various stationary phases geometries. Various analytical strategies have been proposed, such as the use of silica-based monolithic supports, elevated mobile phase temperatures, and columns packed with sub-3 µm superficially porous particles (fused or solid core) or with sub-2 µm fully porous particles for use in ultra-high-pressure LC systems (UHPLC).

12.3.2 Mobile Phases

A mobile phase in reversed-phase chromatograpy consists of mixtures of water or aqueous buffers, to which organic solvents are added, to elute analytes from a reversed-phase column in a selective manner. The added organic solvents must be miscible with water, and the two most common organic solvents used are acetonitrile and methanol. Other solvents can also be used such as ethanol or 2-propanol (isopropyl alcohol) and tetrahydrofuran (THF). The organic solvent is called also a modifier, since it is added to the aqueous solution in the mobile phase in order to modify the polarity of the mobile phase. Water is the most polar solvent in the reversed phase mobile phase; therefore, lowering the polarity of the mobile phase by adding modifiers enhances its elution strength. The two most widely used organic modifiers are acetonitrile and methanol, although acetonitrile is the more popular choice. Isopropanol (2-propanol) can also be used, because of its strong eluting properties, but its use is limited by its high viscosity, which results in higher backpressures. Both acetonitrile and methanol are less viscous than isopropanol, although a mixture of 50:50 percent of methanol:water is also very viscous and causes high backpressures.

All three solvents are essentially UV transparent. This is a crucial property for common reversed phase chromatography since sample components are typically detected by UV detectors. Acetonitrile is more transparent than the others in low UV wavelengths range, therefore it is used almost exclusively when separating molecules with weak or no chromophores (UV-VIS absorbing groups), such as peptides. Most peptides only absorb at low wavelengths in the ultra-violet spectrum (typically less than 225 nm) and acetonitrile provides much lower background absorbance at low wavelengths than the other common solvents.

The pH of the mobile phase can have an important role on the retention of an analyte and can change the selectivity of certain analytes. For samples containing solutes with ionized functional groups, such as amines, carboxyls, phosphates, phosphonates, sulfates, and sulfonates, the ionization of these groups can be controlled using mobile phase buffers.

For example, carboxylic groups in solutes become increasingly negatively charged as the pH of the mobile phase rises above their pKa, hence the whole molecule becomes more polar and less retained on the a-polar stationary phase. In this case, raising the pH of the phase mobile above 4–5 = pH (which is the typical pKa range for carboxylic groups) increases their ionization, hence decreases their retention. Conversely, using a mobile phase at a pH lower than 4 will increase their retention, because it will decrease their ionization degree, rendering them less polar.

The same considerations apply to substances containing basic functional groups, such as amines, whose pKa ranges are around 8 and above, are retained more, as the pH of the mobile phase increases, approaching 8 and above, because they are less ionized, hence less polar. However, in the case of high pH mobile phases, most of the traditional silica gel based Reversed Phase columns are generally limited for use with mobile phases at pH 8 and above, therefore, control over the retention of amines in this range is limited.

The choice of buffer type is an important factor in RP-LC method development, as it can affect the retention, selectivity, and resolution of the analytes of interest.

When selecting a buffer for RP-HPLC, there are a number of factors to consider, including:

- **The Desired pH of the Mobile Phase:** Buffers are most effective around their pKa value, so it is important to choose a buffer with a pKa that is close to the desired mobile phase pH needed.
- **The Solubility of the Buffer in the Organic Solvent:** The buffer must be compatible with the organic solvent that is being used in the mobile phase acetonitrile, methanol, and isopropanol.
- **The UV Cut-Off of the Buffer:** In case of UV detection, the buffer should have a UV absorption that is below the detection wavelength of the analytes of interest. This will prevent the buffer from interfering with the detection of that analytes.
- **The Compatibility of the Buffer with the Detector:** If mass spectrometry (MS) is being used for detection, the buffer must be compatible with the mass spectrometry (MS) instrument. Some buffers, such as those containing phosphate salts, cannot be used with the MS detectors, as they are not volatile as needed, and they interfere with the MS detection by suppressing the analytes ionization, making them undetected by MS.

Some of the most common buffers used in RP-HPLC include:

- **Phosphate Buffers:** Phosphate buffer is versatile and can be used to achieve a wide range of pH values, thanks to 3 pKa values. They also have very low UV background for UV detection. However, they are not appropriate for MS detection.
- **Acetate Buffers:** Acetate buffers are also versatile and can be used to achieve range of pH values typically used in RP-LC. In terms of UV detection at sub 220 nm wavelength, it is not so favorable. The ammonium acetate buffer is compatible with MS.
- **Formate Buffers:** Formate buffers is similar to the acetate buffer in terms of range of pHs used and limited UV detection under 225 nm. Its ammonium acetate is also compatible with MS.
- **Ammonium Buffers:** Ammonium buffers are volatile and are often used in LC-MS methods. They also are limited for low UV detection.

Charged analytes can be separated on a reversed-phase column by the use of ion-pairing (also called ion-interaction). This technique is known as reversed-phase ion-pairing chromatography. Elution can be performed isocratically (the water-solvent composition does not change during the separation process) or by using a solution gradient (the water-solvent composition changes during the separation process, usually by decreasing the polarity).

12.3.3 Applications and Advantages of RPC

Reversed Phase Chromatography is extensively used in proteomics, peptide mapping, and the analysis of post-translational modifications. Its high resolution and sensitivity make it ideal for separating closely related peptides, determining protein sequences, and identifying degradation products. RPC is often coupled with mass spectrometry for precise molecular characterization. In pharmaceutical research, it plays a vital role in the stability testing and formulation analysis of peptide-based drugs.

One of the main advantages of RPC is its robustness and reproducibility. Columns can be operated under high pressures and temperatures, and the use of volatile organic solvents facilitates easy coupling with detection methods such as UV, fluorescence, or mass spectrometry. RPC can also achieve excellent separation of components that are difficult to resolve using other techniques, including isomeric forms and lipid-modified peptides.

12.3.4 Limitations of RPC

However, RPC is not without drawbacks. The use of organic solvents such as acetonitrile can lead to protein unfolding or denaturation, which limits its application in functional protein purification. Moreover, the need for solvent disposal and potential environmental concerns associated with volatile organic compounds are important considerations. Column fouling and limited reusability due to strong retention of hydrophobic compounds can also affect long-term cost-effectiveness.

12.4 COMPARATIVE INSIGHTS BETWEEN HIC AND RPC

Though both HIC and RPC rely on hydrophobic interactions, their operational conditions and outcomes are markedly different. HIC is gentle, maintaining proteins in their active, native states, making it suitable for preparative applications. RPC, while offering superior resolution, operates under denaturing conditions, making it more appropriate for analytical and identification purposes. HIC depends on the manipulation of salt concentrations to elute proteins, whereas RPC uses gradients of organic solvents. These differences mean that HIC is often preferred earlier in the purification process, while RPC is used later, especially when detailed characterization is required.

12.5 RECENT DEVELOPMENTS AND TECHNOLOGICAL ADVANCES

Advances in material science and column technology have significantly enhanced the performance of both HIC and RPC. Modern HIC resins exhibit improved binding capacities, better flow characteristics, and broader pH stability. Innovations in gradient programming and mobile phase optimization have made HIC more adaptable to automated purification systems.

In RPC, developments such as ultra-high performance liquid chromatography (UHPLC) and new stationary phase chemistries have led to shorter run times and higher peak resolution. Hybrid stationary phases that reduce nonspecific binding and improve reproducibility have also been introduced. Furthermore, the integration of microfluidic systems and online detectors has allowed for real-time monitoring and control, making RPC a mainstay in high-throughput analytical workflows.

12.6 CONCLUSION

Reversed Phase and Hydrophobic Interaction Chromatography are two indispensable chromatographic techniques in the realm of bioseparations. Though grounded in similar physicochemical principles, they serve complementary roles across the spectrum of protein and peptide analysis and purification. HIC provides a gentle, preparative approach to separating proteins based on surface hydrophobicity, maintaining biological activity and offering scalability. RPC, in contrast, delivers high-resolution separation ideal for analytical applications and structural characterization, albeit at the cost of denaturation. The choice between these techniques depends on the nature of the biomolecule, the desired outcome of the separation, and the compatibility with downstream applications. As biotechnological and pharmaceutical demands continue to evolve, the refinement and application of HIC and RPC will remain central to advancing the science and engineering of biomolecular separations.

CHAPTER 13

Affinity Chromatography

Affinity chromatography is one of the most powerful and selective techniques used in the field of bioseparations. It is based on the highly specific interaction between a biological molecule and its corresponding ligand. This method harnesses the principle of molecular recognition, where one molecule binds to another with high specificity and affinity, akin to the lock-and-key model. Because of its high selectivity, affinity chromatography is particularly useful in isolating target biomolecules from complex mixtures, such as cell lysates, fermentation broths, and blood plasma, with high purity and yield. Over the years, it has become a critical tool in biochemistry, molecular biology, and pharmaceutical industries for purifying enzymes, antibodies, nucleic acids, and recombinant proteins.

13.1 PRINCIPLE OF AFFINITY CHROMATOGRAPHY

The fundamental principle of affinity chromatography lies in the reversible binding interaction between a specific ligand attached to a stationary matrix and the target molecule present in the mobile phase. The ligand is usually immobilized on a solid support, often composed of agarose, cellulose, or synthetic polymers. When a mixture containing the desired biomolecule is passed through the column, only those molecules with a specific affinity for the ligand bind to the stationary phase. The unbound substances pass through the column, while the target molecule is retained. Subsequently, a change in conditions—such as pH, ionic strength, or the use of a competitive eluent—is introduced to disrupt the interaction, resulting in the elution of the bound molecule in a purified form.

13.2 TYPES OF LIGANDS USED IN AFFINITY CHROMATOGRAPHY

The success of affinity chromatography depends significantly on the choice of ligand, which must have strong and specific binding capacity for the target molecule. Ligands can be broadly categorized into biological and synthetic types. Biological ligands include antibodies, antigens, enzymes, substrates, cofactors, lectins, and nucleic acids. Synthetic ligands, on the other hand, are engineered molecules that mimic natural binding partners and are often more stable and less expensive. For instance, Protein A and Protein G are widely used ligands for purifying immunoglobulins due to their high affinity for the Fc region of antibodies. Other ligands like heparin are employed for purifying growth factors and coagulation proteins, while immobilized metal ions are used in immobilized metal affinity chromatography (IMAC) for purifying histidine-tagged proteins.

13.3 STATIONARY PHASE AND MATRIX DESIGN

The matrix used in affinity chromatography must be inert, stable, and compatible with biological samples. Agarose is one of the most common support materials due to its hydrophilic nature and low non-specific binding. The matrix must also allow for effective ligand coupling, often facilitated by the presence of functional groups like hydroxyls, amines, or carboxyls. These groups can be activated chemically to form stable covalent bonds with the ligand. The pore size and particle diameter of the matrix are also critical as they determine the flow properties and the accessibility of binding sites. Modern resins are designed to minimize non-specific adsorption and to support high flow rates for rapid processing.

13.4 PROCESS OF AFFINITY CHROMATOGRAPHY

The process of affinity chromatography typically involves several steps, starting with the equilibration of the column using a buffer that maintains the activity and structure of both the ligand and the target molecule. The sample is then loaded onto the column, and the target molecule binds to the ligand while other components flow through. This is followed by a wash step to remove any weakly bound or non-specifically bound contaminants. Finally, the elution step is performed by altering the buffer conditions to disrupt the ligand-target interaction. This can be done by changing the pH, ionic strength, or by introducing a competitive ligand that displaces the target molecule. The eluted fractions are then collected and analyzed for purity and yield.

13.5 APPLICATIONS OF AFFINITY CHROMATOGRAPHY

Affinity chromatography is extensively used in both analytical and preparative contexts. In research laboratories, it is used for the purification of biomolecules for structural and functional studies. It is indispensable in proteomics for isolating specific proteins from cell extracts and in genomics for purifying nucleic acids. One of the most significant applications is in the purification of recombinant proteins, especially those with affinity tags like His-tags, GST-tags, or FLAG-tags. These tags are genetically engineered into the protein of interest and allow for rapid and efficient purification using specific ligands.

In the pharmaceutical industry, affinity chromatography plays a vital role in the production of therapeutic proteins, monoclonal antibodies, and vaccines. For example, Protein A chromatography is a standard step in the purification of monoclonal antibodies due to its high selectivity and ability to yield highly pure products. Moreover, affinity chromatography is also used in diagnostic applications, where it helps in the isolation of biomarkers and the development of diagnostic reagents.

13.6 ADVANTAGES OF AFFINITY CHROMATOGRAPHY

The primary advantage of affinity chromatography is its exceptional specificity, which enables the purification of a target molecule in a single step, often achieving high purity and recovery. This selectivity also translates into efficiency, reducing the need for multiple purification steps and thus saving time and resources. The method is highly versatile and can be adapted for a wide range of biomolecules by selecting appropriate ligands. Additionally, the ability to regenerate and reuse affinity columns after proper cleaning and re-equilibration adds to the cost-effectiveness of the technique.

13.7 LIMITATIONS AND CHALLENGES

Despite its many advantages, affinity chromatography has certain limitations. The requirement for a specific ligand, which may be expensive or difficult to produce, can be a significant barrier. In some cases, the interaction between the ligand and the target molecule may be too strong, making elution difficult without affecting the bioactivity of the product. The potential for non-specific binding, although minimized in well-designed systems, can still affect the purity and yield. Furthermore, the method may not be suitable for purifying molecules in large-scale operations if the ligand or matrix is not stable under industrial conditions. Issues related to ligand leaching and column fouling must also be carefully addressed to maintain product consistency and quality.

13.8 INNOVATIONS AND FUTURE DIRECTIONS

Ongoing research and innovation are continuously improving the effectiveness and applicability of affinity chromatography. The development of synthetic ligands and engineered proteins with enhanced stability and binding properties has expanded the range of targets that can be purified. Advances in ligand immobilization techniques are increasing binding efficiency and reducing non-specific interactions. High-throughput formats such as magnetic beads and microfluidic platforms are enabling rapid and miniaturized affinity purification processes, particularly useful in screening and diagnostic applications.

Affinity chromatography is also increasingly being integrated with other chromatographic methods in multi-step purification schemes, combining the selectivity of affinity binding with the resolving power of ion exchange or size exclusion chromatography. In addition, automation and continuous chromatography systems are being adopted in biomanufacturing to improve scalability and throughput while maintaining stringent quality standards.

13.9 CONCLUSION

Affinity chromatography stands as a cornerstone in the field of bioseparations, offering unparalleled specificity and efficiency in the purification of biomolecules. Its reliance on the natural principles of molecular recognition enables it to isolate target compounds from complex biological mixtures with high purity in a single step. Though challenges remain in terms of cost, scalability, and ligand stability, ongoing technological advances continue to enhance its utility and applicability. From academic research to industrial-scale bioprocessing, affinity chromatography remains an indispensable tool that bridges the gap between biological complexity and biochemical precision.

CHAPTER 14

Electrokinetic Methods of Separation

Electrokinetic methods of separation represent a powerful class of analytical techniques that harness electric fields to manipulate and separate charged particles, including ions, molecules, and colloids, in various media. These methods are particularly valuable in the separation of biomolecules, such as nucleic acids, proteins, and peptides, due to their charge-to-mass characteristics. The basic principle underlying electrokinetic separations is the movement of charged species in an electric field, a phenomenon known as electrophoresis. This movement depends on several parameters, including the charge, size, and shape of the molecules, as well as the properties of the supporting medium and the strength of the electric field applied.

14.1 PRINCIPLES OF ELECTROKINETIC SEPARATION

Electrokinetic separations rely on the differential migration of analytes under the influence of an electric field. When an electric field is applied across a medium containing charged particles, those particles begin to move toward the electrode of opposite charge. The velocity at which a molecule moves is known as its electrophoretic mobility, and it is determined by the balance between the electric force acting on the charged particle and the frictional resistance from the medium. Electrophoretic mobility is directly proportional to the charge of the molecule and inversely proportional to its size and the viscosity of the medium.

One of the most common media used for electrokinetic separations is a gel matrix, such as agarose or polyacrylamide, which provides a sieving effect in addition to the influence of the electric field. Molecules not only migrate based on their charge but are also impeded based on their size, with larger molecules facing more resistance. This dual effect allows for a high-resolution separation of complex mixtures, making electrokinetic techniques extremely useful in biochemical and molecular biology research.

14.2 CAPILLARY ELECTROPHORESIS

Capillary electrophoresis (CE) is a modern and highly efficient electrokinetic technique that uses narrow-bore capillaries to achieve high-resolution separation of analytes in a liquid phase. It operates on the principle of differential electrophoretic mobility and electroosmotic flow (EOF), the latter being a phenomenon where the bulk liquid inside the capillary also moves due to the interaction between the electric field and the charged capillary walls. The combination of these two effects enables the rapid and effective separation of both small and large molecules.

In capillary electrophoresis, the sample is introduced into the capillary and an electric field is applied across it. The analytes migrate at different speeds based on their charge-to-size ratios, while the electroosmotic flow pushes all components toward the detector end. This results in well-defined peaks and highly efficient separations. CE is particularly advantageous due to its high resolution, minimal sample and reagent requirements, and compatibility with a variety of detection systems such as UV-Vis, fluorescence, and mass spectrometry. It is widely used in pharmaceutical analysis, forensic science, and clinical diagnostics.

14.3 GEL ELECTROPHORESIS

Gel electrophoresis is a classic and widely used electrokinetic technique, especially in the analysis of nucleic acids and proteins. It involves the use of a gel matrix, typically agarose for DNA and RNA, or polyacrylamide for proteins, cast into a slab or tube form. The gel acts as a molecular sieve, allowing smaller molecules to move faster than larger ones under the influence of an electric field. Samples are loaded into wells within the gel, and upon applying voltage, the charged molecules migrate toward the anode or cathode depending on their net charge.

One of the key advantages of gel electrophoresis is its simplicity and effectiveness in visualizing molecular size and purity. After separation, the molecules can be stained with dyes such as ethidium bromide for nucleic acids or Coomassie Brilliant Blue for proteins, and then visualized using UV or visible light.

Variants of gel electrophoresis, such as SDS-PAGE (sodium dodecyl sulfate polyacrylamide gel electrophoresis), allow for the separation of proteins based purely on molecular weight by denaturing them and imparting a uniform negative charge.

14.3.1 Physical

Electrophoresis is a process that enables the sorting of molecules based on charge, size, or shape. Using an electric field, molecules such as DNA can be made to move through a gel made of agarose or polyacrylamide. The electric field consists of a negative charge at one end which pushes the molecules through the gel and a positive charge at the other end that pulls the molecules through the gel. The molecules being sorted are dispensed into a well in the gel material. The gel is placed in an electrophoresis chamber, which is then connected to a power source. When the electric field is applied, the larger molecules move more slowly through the gel while the smaller molecules move faster. The different sized molecules form distinct bands on the gel.

The term "gel" in this instance refers to the matrix used to contain, then separate the target molecules. In most cases, the gel is a crosslinked polymer whose composition and porosity are chosen based on the specific weight and composition of the target to be analyzed. When separating proteins or small nucleic acids (DNA, RNA, or oligonucleotides), the gel is usually composed of different concentrations of acrylamide and a cross-linker, producing different sized mesh networks of polyacrylamide. When separating larger nucleic acids (greater than a few hundred bases), the preferred matrix is purified agarose. In both cases, the gel forms a solid yet porous matrix. Acrylamide, in contrast to polyacrylamide, is a neurotoxin and must be handled using appropriate safety precautions to avoid poisoning. Agarose is composed of long unbranched chains of uncharged carbohydrates without cross-links, resulting in a gel with large pores allowing for the separation of macromolecules and macromolecular complexes.

Electrophoresis refers to the electromotive force (EMF) that is used to move the molecules through the gel matrix. By placing the molecules in wells in the gel and applying an electric field, the molecules will move through the matrix at different rates, determined largely by their mass when the charge-to-mass ratio (Z) of all species is uniform. However, when charges are not uniform, the electrical field generated by the electrophoresis procedure will cause the molecules to migrate differentially according to charge. Species that are net positively charged will migrate towards the cathode (which is negatively charged because this is an electrolytic rather than galvanic cell), whereas species that are net negatively charged will migrate towards the positively charged anode. Mass remains a factor in the speed with which these non-uniformly charged molecules migrate through the matrix toward their respective electrodes.

If several samples have been loaded into adjacent wells in the gel, they will run parallel in individual lanes. Depending on the number of different molecules, each lane shows the separation of the components from the original mixture as one or more distinct bands, one band per component. Incomplete separation of the components can lead to overlapping bands, or indistinguishable smears representing multiple unresolved components. Bands in different lanes that end up at the same distance from the top contain molecules that passed through the gel at the same speed, which usually means they are approximately the same size. There are molecular weight size markers available that contain a mixture of molecules of known sizes. If such a marker was run on one lane in the gel parallel to the unknown samples, the bands observed can be compared to those of the unknown to determine their size. The distance a band travels is approximately inversely proportional to the logarithm of the size of the molecule. (Equivalently, the distance traveled is inversely proportional to the log of the samples's molecular weight).

There are limits to electrophoretic techniques. Since passing a current through a gel causes heating, gels may melt during electrophoresis. Electrophoresis is performed in buffer solutions to reduce pH changes due to the electric field, which is important because the charge of DNA and RNA depends on pH, but running for too long can exhaust the buffering capacity of the solution.

There are also limitations in determining the molecular weight by SDS-PAGE, especially when trying to find the MW of an unknown protein. Certain biological variables are difficult or impossible to minimize and can affect electrophoretic migration. Such factors include protein structure, post-translational modifications, and amino acid composition.

For example, tropomyosin is an acidic protein that migrates abnormally on SDS-PAGE gels. This is because the acidic residues are repelled by the negatively charged SDS, leading to an inaccurate mass-to-charge ratio and migration. Further, different preparations of genetic material may not migrate consistently with each other, for morphological or other reasons.

14.3.2 Types of Gel

The types of gel most typically used are agarose and polyacrylamide gels. Each type of gel is well-suited to different types and sizes of the analyte. Polyacrylamide gels are usually used for proteins and have very high resolving power for small fragments of DNA (5-500 bp). Agarose gels, on the other hand, have lower resolving power for DNA but a greater range of separation, and are therefore usually used for DNA fragments of 50–20,000 bp in size. (The resolution of over 6 Mb is possible with pulsed field gel electrophoresis (PFGE).) Polyacrylamide gels are run in a vertical configuration while agarose gels are typically run horizontally in a submarine mode. They also differ in their casting methodology, as agarose sets thermally, while polyacrylamide forms in a chemical polymerization reaction.

Agarose

Agarose gels are made from the natural polysaccharide polymers extracted from seaweed. Agarose gels are easily cast and handled compared to other matrices because the gel setting is a physical rather than chemical change. Samples are also easily recovered. After the experiment is finished, the resulting gel can be stored in a plastic bag in a refrigerator.

Agarose gels do not have a uniform pore size, but are optimal for electrophoresis of proteins that are larger than 200 kDa. Agarose gel electrophoresis can also be used for the separation of DNA fragments ranging from 50 base pair to several megabases (millions of bases), the largest of which require specialized apparatus. The distance between DNA bands of different lengths is influenced by the percent agarose in the gel, with higher percentages requiring longer run times, sometimes days. Instead high percentage agarose gels should be run with a pulsed field electrophoresis (PFE), or field inversion electrophoresis.

"Most agarose gels are made with between 0.7% (good separation or resolution of large 5–10kb DNA fragments) and 2% (good resolution for small 0.2–1kb fragments) agarose dissolved in electrophoresis buffer. Up to 3% can be used for separating very tiny fragments but a vertical polyacrylamide gel is more appropriate in this case. Low percentage gels are very weak and may break when you try to lift them. High percentage gels are often brittle and do not set evenly. 1% gels are common for many applications."

Polyacrylamide

Polyacrylamide gel electrophoresis (PAGE) is used for separating proteins ranging in size from 5 to 2,000 kDa due to the uniform pore size provided by the polyacrylamide gel. Pore size is controlled by modulating the concentrations of acrylamide and bis-acrylamide powder and by the polymerization time used in creating a gel. Care must be used when creating this type of gel, as acrylamide is a potent neurotoxin in its liquid and powdered forms.

Traditional DNA sequencing techniques such as Maxam-Gilbert or Sanger methods used polyacrylamide gels to separate DNA fragments differing by a single base-pair in length so the sequence could be read. Most modern DNA separation methods now use agarose gels, except for particularly small DNA fragments. It is currently most often used in the field of immunology and protein analysis, often used to separate different proteins or isoforms of the same protein into separate bands. These can be transferred onto a nitrocellulose or PVDF membrane to be probed with antibodies and corresponding markers, such as in a western blot.

Typically resolving gels are made in 6%, 8%, 10%, 12% or 15%. Stacking gel (5%) is poured on top of the resolving gel and a gel comb (which forms the wells and defines the lanes where proteins, sample buffer, and ladders will be placed) is inserted. The percentage chosen depends on the size of the protein that one wishes to identify or probe in the sample. The smaller the known weight, the higher the percentage that should be used. Changes in the buffer system of the gel can help to further resolve proteins of very small sizes.

Starch

Partially hydrolysed potato starch makes for another non-toxic medium for protein electrophoresis. The gels are slightly more opaque than acrylamide or agarose. Non-denatured proteins can be separated according to charge and size. They are visualised using Napthal Black or Amido Black staining. Typical starch gel concentrations are 5% to 10%.

14.3.3 Gel Conditions

Denaturing

Denaturing gels are run under conditions that disrupt the natural structure of the analyte, causing it to unfold into a linear chain. Thus, the mobility of each macromolecule depends only on its linear length and its mass-to-charge ratio. Thus, the secondary, tertiary, and quaternary levels of biomolecular structure are disrupted, leaving only the primary structure to be analyzed.

Nucleic acids are often denatured by including urea in the buffer, while proteins are denatured using sodium dodecyl sulfate, usually as part of the SDS-PAGE process. For full denaturation of proteins, it is also necessary to reduce the covalent disulfide bonds that stabilize their tertiary and quaternary structure, a method called reducing PAGE. Reducing conditions are usually maintained by the addition of beta-mercaptoethanol or dithiothreitol. For a general analysis of protein samples, reducing PAGE is the most common form of protein electrophoresis.

Denaturing conditions are necessary for proper estimation of molecular weight of RNA. RNA is able to form more intramolecular interactions than DNA which may result in change of its electrophoretic mobility. Urea, DMSO and glyoxal are the most often used denaturing agents to disrupt RNA structure. Originally, highly toxic methylmercury hydroxide was often used in denaturing RNA electrophoresis, but it may be method of choice for some samples.

Denaturing gel electrophoresis is used in the DNA and RNA banding pattern-based methods temperature gradient gel electrophoresis (TGGE) and denaturing gradient gel electrophoresis (DGGE).

Native

Native gels are run in non-denaturing conditions so that the analyte's natural structure is maintained. This allows the physical size of the folded or assembled complex to affect the mobility, allowing for analysis of all four levels of the biomolecular structure. For biological samples, detergents are used only to the extent that they are necessary to lyse lipid membranes in the cell. Complexes remain—for the most part—associated and folded as they would be in the cell. One downside, however, is that complexes may not separate cleanly or predictably, as it is difficult to predict how the molecule's shape and size will affect its mobility. These effects have been addressed by preparative native PAGE. Unlike denaturing methods, native gel electrophoresis does not use a charged denaturing agent. The molecules being separated (usually proteins or nucleic acids) therefore differ not only in molecular mass and intrinsic charge, but also the cross-sectional area, and thus experience different electrophoretic forces dependent on the shape of the overall structure. For proteins, since they remain in the native state they may be visualized not only by general protein staining reagents but also by specific enzyme-linked staining.

A specific experiment example of an application of native gel electrophoresis is to check for enzymatic activity to verify the presence of the enzyme in the sample during protein purification. For example, for the protein alkaline phosphatase, the staining solution is a mixture of 4-chloro-2-2methylbenzenediazonium salt with 3-phospho-2-naphthoic acid-2'-4'-dimethyl aniline in Tris buffer. This stain is commercially sold as a kit for staining gels. If the protein is present, the mechanism of the reaction takes place in the following order: it starts with the de-phosphorylation of 3-phospho-2-naphthoic acid-2'-4'-dimethyl aniline by alkaline phosphatase (water is needed for the reaction). The phosphate group is released and replaced by an alcohol group from water. The electrophile 4- chloro-2-2 methylbenzenediazonium (Fast Red TR Diazonium salt) displaces the alcohol group forming the final product Red Azo dye. As its name implies, this is the final visible-red product of the reaction. In undergraduate academic experimentation of protein purification, the gel is usually run next to commercial purified samples to visualize the results and conclude whether or not purification was successful.

Native gel electrophoresis is typically used in proteomics and metallomics. However, native PAGE is also used to scan genes (DNA) for unknown mutations as in single-strand conformation polymorphism.

14.3.4 Buffers

Buffers in gel electrophoresis are used to provide ions that carry a current and to maintain the pH at a relatively constant value. These buffers have plenty of ions in them, which is necessary for the passage of electricity through them. Something like distilled water or benzene contains few ions, which is not ideal for the use in electrophoresis. There are a number of buffers used for electrophoresis. The most common being, for nucleic acids Tris/Acetate/EDTA (TAE), Tris/Borate/EDTA (TBE). Many other buffers have been proposed, e.g. lithium borate, which is rarely used, based on Pubmed citations (LB), isoelectric histidine, pK matched goods buffers, etc.; in most cases the purported rationale is lower current (less heat) matched ion mobilities, which leads to longer buffer life. Borate is problematic; Borate can polymerize, or interact with cis diols such as those found in RNA. TAE has the lowest buffering capacity but provides the best resolution for larger DNA. This means a lower voltage and more time, but a better product. LB is relatively new and is ineffective in resolving fragments larger than 5 kbp; However, with its low conductivity, a much higher voltage could be used (up to 35 V/cm), which means a shorter analysis time for routine electrophoresis. As low as one base pair size difference could be resolved in 3% agarose gel with an extremely low conductivity medium (1 mM Lithium borate).

Most SDS-PAGE protein separations are performed using a "discontinuous" (or DISC) buffer system that significantly enhances the sharpness of the bands within the gel. During electrophoresis in a discontinuous gel system, an ion gradient is formed in the early stage of electrophoresis that causes all of the proteins to focus on a single sharp band in a process called isotachophoresis. Separation of the proteins by size is achieved in the lower, "resolving" region of the gel. The resolving gel typically has a much smaller pore size, which leads to a sieving effect that now determines the electrophoretic mobility of the proteins.

14.3.5 Visualization

After the electrophoresis is complete, the molecules in the gel can be stained to make them visible. DNA may be visualized using ethidium bromide which, when intercalated into DNA, fluoresce under ultraviolet light, while protein may be visualised using silver stain or Coomassie brilliant blue dye.

Other methods may also be used to visualize the separation of the mixture's components on the gel. If the molecules to be separated contain radioactivity, for example in a DNA sequencing gel, an autoradiogram can be recorded of the gel. Photographs can be taken of gels, often using a gel doc system. Gels are then commonly labelled for presentation and scientific records on the popular figure-creation website, SciUGo.

14.3.6 Downstream Processing

After separation, an additional separation method may then be used, such as isoelectric focusing or SDS-PAGE. The gel will then be physically cut, and the protein complexes extracted from each portion separately. Each extract may then be analysed, such as by peptide mass fingerprinting or de novo peptide sequencing after in-gel digestion. This can provide a great deal of information about the identities of the proteins in a complex.

14.3.7 Applications

- Estimation of the size of DNA molecules following restriction enzyme digestion, e.g. in restriction mapping of cloned DNA.
- Analysis of PCR products, e.g. in molecular genetic diagnosis or genetic fingerprinting
- Separation of restricted genomic DNA prior to Southern transfer, or of RNA prior to Northern transfer.

Gel electrophoresis is used in forensics, molecular biology, genetics, microbiology and biochemistry. The results can be analyzed quantitatively by visualizing the gel with UV light and a gel imaging device. The image is recorded with a computer-operated camera, and the intensity of the band or spot of interest is measured and compared against standard or markers loaded on the same gel. The measurement and analysis are mostly done with specialized software.

Depending on the type of analysis being performed, other techniques are often implemented in conjunction with the results of gel electrophoresis, providing a wide range of field-specific applications.

Nucleic Acids

In the case of nucleic acids, the direction of migration, from negative to positive electrodes, is due to the naturally occurring negative charge carried by their sugar-phosphate backbone.

Double-stranded DNA fragments naturally behave as long rods, so their migration through the gel is relative to their size or, for cyclic fragments, their radius of gyration. Circular DNA such as plasmids, however, may show multiple bands, the speed of migration may depend on whether it is relaxed or supercoiled. Single-stranded DNA or RNA tends to fold up into molecules with complex shapes and migrate through the gel in a complicated manner based on their tertiary structure. Therefore, agents that disrupt the hydrogen bonds, such as sodium hydroxide or formamide, are used to denature the nucleic acids and cause them to behave as long rods again.

Electrophoresis of RNA samples can be used to check for genomic DNA contamination and also for RNA degradation. RNA from eukaryotic organisms shows distinct bands of 28s and 18s rRNA, the 28s band being approximately twice as intense as the 18s band. Degraded RNA has less sharply defined bands, has a smeared appearance, and the intensity ratio is less than 2:1.

Proteins

Proteins, unlike nucleic acids, can have varying charges and complex shapes, therefore they may not migrate into the polyacrylamide gel at similar rates, or all when placing a negative to positive EMF on the sample. Proteins, therefore, are usually denatured in the presence of a detergent such as sodium dodecyl sulfate (SDS) that coats the proteins with a negative charge.

Generally, the amount of SDS bound is relative to the size of the protein (usually 1.4g SDS per gram of protein), so that the resulting denatured proteins have an overall negative charge, and all the proteins have a similar charge-to-mass ratio. Since denatured proteins act like long rods instead of having a complex tertiary shape, the rate at which the resulting SDS coated proteins migrate in the gel is relative only to their size and not their charge or shape.

Proteins are usually analyzed by sodium dodecyl sulfate polyacrylamide gel electrophoresis (SDS-PAGE), by native gel electrophoresis, by preparative native gel electrophoresis (QPNC-PAGE), or by 2-D electrophoresis.

Characterization through ligand interaction may be performed by electroblotting or by affinity electrophoresis in agarose or by capillary electrophoresis as for estimation of binding constants and determination of structural features like glycan content through lectin binding.

Nanoparticles

A novel application for gel electrophoresis is the separation or characterization of metal or metal oxide nanoparticles (e.g. Au, Ag, ZnO, SiO2) regarding the size, shape, or surface chemistry of the nanoparticles. The scope is to obtain a more homogeneous sample (e.g. narrower particle size distribution), which then can be used in further products/processes (e.g. self-assembly processes). For the separation of nanoparticles within a gel, the key parameter is the ratio of the particle size to the mesh size, whereby two migration mechanisms were identified: the unrestricted mechanism, where the particle size << mesh size, and the restricted mechanism, where particle size is similar to mesh size.

14.4 ISOELECTRIC FOCUSING

Isoelectric focusing (IEF) is a specialized electrokinetic technique used to separate amphoteric molecules like proteins based on their isoelectric point (pI), the pH at which the molecule carries no net charge. In IEF, a pH gradient is established within a gel, and when an electric field is applied, proteins migrate through the gradient until they reach the pH corresponding to their pI. At this point, they stop moving because the net charge is zero, resulting in focused, sharp bands.

This method provides extremely high resolution and is often used as the first dimension in two-dimensional gel electrophoresis, a technique that combines IEF with SDS-PAGE to analyze complex protein mixtures. IEF is invaluable in proteomics, diagnostics, and the characterization of protein isoforms and post-translational modifications. Its main strength lies in its ability to resolve proteins that differ by very small pI values, often less than 0.01 units.

14.5 MICELLAR ELECTROKINETIC CHROMATOGRAPHY

Micellar electrokinetic chromatography (MEKC) is an extension of capillary electrophoresis that allows for the separation of both charged and neutral analytes. In MEKC, surfactants are added to the buffer solution, forming micelles that act as a pseudo-stationary phase. Neutral molecules partition between the micelles and the aqueous phase based on their hydrophobicity, while charged molecules are also influenced by their electrophoretic mobility.

MEKC is particularly useful for separating mixtures of compounds that are otherwise difficult to resolve by traditional CE, especially small organic molecules, drugs, and peptides. It retains the advantages of CE, such as high efficiency and low sample consumption, while expanding the range of analytes that can be effectively separated.

14.6 APPLICATIONS OF ELECTROKINETIC SEPARATIONS

Electrokinetic separation techniques have found widespread application in various fields due to their high resolution, speed, and low reagent consumption. In the field of molecular biology, they are indispensable for DNA and RNA analysis, such as in genotyping, sequencing, and mutation detection. Protein analysis, including the assessment of purity, molecular weight, and charge heterogeneity, is another major application area.

In clinical laboratories, electrokinetic methods are used for diagnostic purposes, including the separation and identification of hemoglobin variants and serum proteins. Forensic science employs these techniques in DNA fingerprinting and toxicological screening, while pharmaceutical companies use them in drug development, quality control, and pharmacokinetic studies.

14.7 ADVANTAGES OF ELECTROKINETIC METHODS

The main advantages of electrokinetic methods include their high separation efficiency, rapid analysis times, and minimal requirements for sample and buffer volumes. These techniques are also highly adaptable, allowing for the analysis of a broad range of analytes, from small ions to large biomolecules. The ability to couple electrokinetic separations with powerful detectors like mass spectrometry enhances their analytical capabilities, enabling both qualitative and quantitative analysis.

Additionally, many electrokinetic methods are suitable for miniaturization and automation, making them ideal for high-throughput applications and integration into lab-on-a-chip devices. This has opened new avenues in fields such as point-of-care diagnostics and personalized medicine, where rapid and accurate analyses are crucial.

14.8 LIMITATIONS AND CHALLENGES

Despite their numerous advantages, electrokinetic separation techniques are not without limitations. One significant challenge is the sensitivity to matrix effects, as impurities or high salt concentrations in samples can interfere with migration patterns and resolution. Another issue is the relatively low loading capacity compared to traditional chromatographic methods, which can limit their use in preparative-scale separations.

Moreover, the reproducibility of certain methods, especially those involving gel matrices, can vary due to differences in gel composition and preparation. Temperature control is also critical, as Joule heating from the applied electric field can affect the migration and resolution. Advanced instrumentation and careful optimization of experimental conditions are often necessary to overcome these limitations.

14.9 CONCLUSION

Electrokinetic methods of separation have revolutionized analytical science by providing powerful tools for the rapid and efficient analysis of charged species. From classical gel electrophoresis to advanced capillary electrophoresis and micellar electrokinetic chromatography, these techniques offer high precision, sensitivity, and versatility across a wide range of applications. Despite some technical challenges, ongoing advances in materials, instrumentation, and automation continue to enhance their performance and expand their utility. As analytical demands grow in complexity and sensitivity, electrokinetic separation methods are poised to remain at the forefront of scientific and technological innovation.

CHAPTER 15

Finishing Operations and Formulation

Finishing operations and formulation represent the final stages in the production of pharmaceutical, biotechnological, and chemical products, where the active ingredients are transformed into a final dosage form that is stable, efficacious, safe, and convenient for administration. These stages are crucial for ensuring the therapeutic effectiveness of the product and maintaining its quality during storage and transportation. Formulation involves the careful combination of active pharmaceutical ingredients (APIs) with suitable excipients, while finishing operations refer to the downstream processes that prepare the final product for packaging and delivery.

15.1 IMPORTANCE OF FORMULATION IN PRODUCT DEVELOPMENT

Formulation is the science and art of designing a product that delivers the active component in a desired manner to the site of action. It involves converting the raw drug substance into a usable dosage form such as tablets, capsules, injectables, or topical preparations. The process starts with understanding the physical and chemical properties of the API, including solubility, stability, and bioavailability. These properties influence the selection of excipients, which are inert substances added to aid in processing, stability, and delivery of the drug. Excipients may serve various roles such as binders, disintegrants, lubricants, fillers, preservatives, and stabilizers. A successful formulation ensures that the drug is released at the right time, in the right place, and at the right rate in the body. It also addresses challenges like improving solubility of poorly water-soluble drugs, enhancing bioavailability, masking unpleasant tastes, and prolonging shelf life. Formulation scientists must consider regulatory requirements, patient compliance, and cost-effectiveness while designing a formulation. Different drug delivery systems—such as sustained release, controlled release, or targeted delivery—are developed depending on the therapeutic goals.

15.2 SOLID DOSAGE FORMS AND THEIR FINISHING PROCESSES

Solid dosage forms, such as tablets and capsules, are among the most common forms of pharmaceutical products due to their ease of administration, accurate dosing, and good stability. The manufacturing of these forms includes blending, granulation, drying, compression, and coating, followed by finishing operations like polishing, printing, and packaging. Blending ensures uniform distribution of API and excipients, while granulation improves flowability and compressibility. Compression transforms the granules into tablets using specialized machinery.

Finishing operations in solid dosage form production involve processes like coating, which improves the appearance, masks taste or odor, and controls drug release. Coating can be done using sugar, film, or enteric materials depending on the desired function. After coating, tablets may be polished or deburred to remove sharp edges. Printing is also an essential finishing step used for branding, identification, and regulatory compliance. Finally, tablets or capsules are filled into suitable containers—blister packs, bottles, or strip packs—under controlled environments to maintain product integrity and prevent contamination.

15.3 LIQUID DOSAGE FORMS AND THEIR FINALIZATION

Liquid dosage forms, including solutions, suspensions, and emulsions, are typically used for pediatric, geriatric, or intravenous administration. The formulation of these products involves dissolution or dispersion of the API in a suitable solvent, along with excipients that ensure stability, taste masking, and preservation. The formulation must ensure the homogeneity of the API throughout the container to guarantee consistent dosing.

Finishing operations for liquid formulations include filtration to remove particulates and microbial contaminants, followed by filling into bottles or ampoules under aseptic conditions. The filled containers are then sealed, labeled, and inspected for defects. The choice of container material—glass or plastic—depends on the chemical compatibility with the formulation. Labeling must include information such as dosage, storage conditions, expiry date, and batch number, all of which are critical for safe and effective use.

15.4 PARENTERAL FORMULATIONS AND ASEPTIC PROCESSING

Parenteral formulations are sterile preparations intended for administration through injection. They are typically solutions, suspensions, or emulsions, and are highly sensitive to contamination and degradation. The formulation process for parenterals requires rigorous attention to sterility, isotonicity, pH adjustment, and the use of pyrogen-free water. Suitable buffers and stabilizers are often included to maintain product integrity.

Finishing operations for parenteral products are conducted in sterile environments using highly specialized equipment. The solution is first filtered and filled into ampoules, vials, or prefilled syringes using aseptic techniques. These are then sealed—either by heat fusion or rubber closures—and sterilized if the product is heat stable. Labeling and packaging follow strict regulatory guidelines to ensure product safety. The integrity of the container and seal is tested to avoid microbial ingress during storage.

15.5 TOPICAL AND SEMI-SOLID DOSAGE FORMS

Topical and semi-solid formulations, such as creams, ointments, gels, and lotions, are designed for local application to the skin or mucous membranes. These dosage forms require formulation strategies that ensure appropriate viscosity, spreadability, absorption, and stability. The choice of base—hydrophilic or lipophilic—greatly affects the drug release profile and patient acceptability.

Finishing operations for topical products involve homogenization to ensure consistency, followed by filling into tubes, jars, or pump dispensers. The equipment used must ensure accurate dosing and prevent contamination during filling. Sealing is typically done by crimping or induction sealing to preserve the product's integrity. Labeling must indicate the application area, method of use, and storage instructions to avoid misuse.

15.6 INHALATION AND TRANSDERMAL DOSAGE FORMS

Inhalation products, such as metered-dose inhalers (MDIs), dry powder inhalers (DPIs), and nebulizers, deliver medication directly to the lungs for rapid action. Formulation considerations include particle size control, moisture protection, and propellant compatibility. Transdermal systems, on the other hand, involve the formulation of drugs into patches that provide controlled drug release through the skin over extended periods.

The finishing operations for these advanced delivery systems are highly specialized. Inhalers are assembled under precise conditions to ensure accurate dosing per actuation and proper aerosolization. Transdermal patches require uniform drug distribution in the adhesive matrix, precision cutting, and lamination before packaging. In both cases, packaging is designed to protect the product from environmental factors and ensure ease of use.

15.7 QUALITY CONTROL AND REGULATORY COMPLIANCE

At every stage of formulation and finishing operations, stringent quality control measures are implemented to ensure compliance with Good Manufacturing Practices (GMP) and regulatory standards. This includes in-process checks for weight variation, hardness, content uniformity, and microbial limits. Stability testing is conducted under various conditions to predict shelf life and storage requirements. Regulatory authorities such as the FDA, EMA, and national drug control organizations have established guidelines that manufacturers must follow. These include documentation of raw materials, manufacturing procedures, quality control results, and batch records. Any deviation from the standard procedure must be documented, investigated, and rectified to prevent compromise in product quality.

15.8 PACKAGING AND STORAGE CONSIDERATIONS

Packaging is a vital finishing operation that not only contains and protects the product but also provides information for proper usage. Primary packaging materials come in direct contact with the product and must be inert, impermeable, and compatible. Secondary packaging offers additional protection and facilitates transportation. Tamper-evident and child-resistant features are commonly incorporated to enhance safety.

Storage conditions are clearly defined on the product label and are based on stability data. These conditions must be maintained during transportation and warehousing to prevent degradation. Temperature-controlled storage is essential for sensitive products like vaccines and biologics.

15.9 CONCLUSION

Finishing operations and formulation are integral components of pharmaceutical manufacturing, ensuring that the final product is effective, safe, and user-friendly. From the selection of excipients and dosage form design to final packaging and labeling, every step requires careful planning, precise execution, and strict adherence to regulatory standards. Innovations in drug delivery systems and continuous improvement in manufacturing technologies are further advancing the scope of formulation science and finishing operations. As the demand for personalized and high-quality healthcare products grows, the role of these processes becomes even more significant in delivering therapeutic benefits to patients worldwide.

Bibliography

A Century of Separation Science Editor: Haleem J. Issaq Publisher: CRC Press Year: 2001

Biochemical Engineering Fundamentals by James E. Bailey and David F. Ollis; McGraw-Hill, 2nd Edition, 1986.

Bioprocess Engineering Principles by Pauline M. Doran; Academic Press, 2nd Edition, 2012.

Bioprocess Engineering: Basic Concepts by Michael L. Shuler and Fikret Kargi; Prentice Hall, 2nd Edition, 2001.

Bioseparation Process Science by Henry G. Schwartzberg and Richard A. Wallace; Academic Press, 1990.

Bioseparations and Bioprocessing: A Handbook edited by Gunter Jagschies et al.; Wiley-VCH, 2007.

Bioseparations Downstream Processing for Biotechnology by Paul A. Belter, E.L. Cussler, and Wei-Shou Hu; Wiley-Interscience, 1988.

Bioseparations Engineering: Principles, Practice and Economics by Michael R. Ladisch; Wiley-Interscience, 2001.

Bioseparations Science and Engineering by Roger G. Harrison, Paul W. Todd, Scott R. Rudge, and Demetri P. Petrides; Oxford University Press, 2003.

Bioseparations: Principles and Techniques by B. Sivasankar; Prentice Hall India, 2005.

Biotechnology: A Textbook of Industrial Microbiology by Wulf Crueger and Anneliese Crueger; Sinauer Associates, 2nd Edition, 1990.

Capillary Electrophoresis: Principles and Practice Author: Reinhard Kuhn Publisher: Springer Year: 1993

Chromatographic Analysis of Pharmaceuticals Authors: John A. Adamovics, John W. Dolan Publisher: CRC Press Year: 2005

Chromatographic Methods Author: A. Braithwaite Publisher: Springer Year: 1995

Chromatographic Methods in Biotechnology by A. M. A. M. El-Mansi and C. F. A. Bryce; CRC Press, 2007.

Chromatography and Separation Science Author: Satinder Ahuja Publisher: Academic Press Year: 2002

Chromatography Editor: Ivor Smith Publisher: Butterworth-Heinemann Year: 1969

Chromatography Today Author: Colin F. Poole Publisher: Elsevier Year: 2012

Chromatography: A Science of Discovery Author: Robert L. Wixom Publisher: Wiley Year: 2011

Chromatography: Concepts and Contrasts Author: James M. Miller Publisher: Wiley Year: 2009

Chromatography: Fundamentals and Applications of Chromatography and Related Differential Migration Methods – Part B: Applications Authors: E. Heftmann, G. Zweig Publisher: Elsevier Year: 2004

CRC Handbook of Chromatography: Analysis of Lipids Authors: Kumar D. Mukherjee, Nikolaus Weber Publisher: CRC Press Year: 1993

Downstream Industrial Biotechnology: Recovery and Purification edited by Michael C. Flickinger; Wiley, 2013.

Downstream Processing in Biotechnology: A Practical Approach edited by P.A. Belter and E.L. Cussler; Wiley-Interscience, 1988.

Downstream Processing of Proteins: Methods and Protocols edited by Paul Cutler; Humana Press, 2004.

Encyclopedia of Chromatography, Third Edition Editor: Jack Cazes Publisher: CRC Press Year: 2009

Gas Chromatography and Mass Spectrometry: A Practical Guide Authors: O. David Sparkman, Zelda Penton, Fulton G. Kitson Publisher: Academic Press Year: 2011

Gas Chromatography and Mass Spectrometry: Theory and Practice Authors: James M. Miller, Robert E. White Publisher: Wiley Year: 2003

Gas Chromatography: Principles, Techniques and Applications Author: Alan J. Handley Publisher: Wiley Year: 2001

Guide to Gas Chromatography Literature Author: Austin V. Signeur Publisher: Springer Year: 1964

Handbook of Bioseparations edited by Satinder Ahuja; Academic Press, 2000.

Handbook of Chromatography: Liquid Chromatography of Polycyclic Aromatic Hydrocarbons Author: Hamir S. Rathore Publisher: CRC Press Year: 1993

Handbook of HPLC Editor: Danilo Corradini Publisher: CRC Press Year: 2005

Handbook of Separation Process Technology by Ronald W. Rousseau; Wiley-Interscience, 1987.

Handbook of Thin-Layer Chromatography Editors: Joseph Sherma, Bernard Fried Publisher: CRC Press Year: 2003

High-Performance Liquid Chromatography: Advances and Perspectives Editor: Csaba Horváth Publisher: Academic Press Year: 1980

High-Performance Liquid Chromatography: Fundamental Principles and Practice Author: W. M. A. Niessen Publisher: Elsevier Year: 2001

Introduction to Chromatography Authors: J. M. Bobbitt, A. E. Schwarting, R. J. Gritter Publisher: Reinhold Publishing Co. Year: 1968

Introduction to Modern Liquid Chromatography Authors: Lloyd R. Snyder, Joseph J. Kirkland, John W. Dolan Publisher: Wiley Year: 2010

Ion Chromatography Author: Hamish Small Publisher: Plenum Press Year: 1989

Liquid Chromatography: Fundamentals and Instrumentation Editors: Salvatore Fanali, Paul R. Haddad, Colin Poole, Marja-Liisa Riekkola Publisher: Elsevier Year: 2013

Membrane Separations in Biotechnology by William B. Krantz and A. L. Zydney; CRC Press, 1996.

Membrane Technology and Applications by Richard W. Baker; Wiley, 3rd Edition, 2012.

Modern Practice of Gas Chromatography Editor: Robert L. Grob Publisher: Wiley-Interscience Year: 2004

Practical Gas Chromatography: A Comprehensive Reference Editor: Katja Dettmer-Wilde Publisher: Springer Year: 2014

Practical HPLC Method Development Authors: Lloyd R. Snyder, Joseph J. Kirkland, John W. Dolan Publisher: Wiley Year: 1997

Principles of Bioseparations Engineering by Raja Ghosh; World Scientific Publishing, 2006.

Principles of Protein Purification by Richard J. Simpson; Springer, 3rd Edition, 2010.

Protein Chromatography: Process Development and Scale-Up edited by Giorgio Carta and Alois Jungbauer; Wiley-VCH, 2010.

Protein Purification: Design and Scale Up of Downstream Processing by Peter A. Belter and E.L. Cussler; Wiley-Interscience, 1988.

Protein Purification: Principles and Practice by Robert K. Scopes; Springer, 3rd Edition, 1994.

Protein Purification: Principles, High Resolution Methods, and Applications by Jan-Christer Janson; Wiley, 3rd Edition, 2011.

Separation Process Engineering: Includes Mass Transfer Analysis by Phillip C. Wankat; Prentice Hall, 3rd Edition, 2012.

Separation Process Principles by J.D. Seader, Ernest J. Henley, and D. Keith Roper; Wiley, 3rd Edition, 2010.

Supercritical Fluid Chromatography: Techniques and Applications Editor: Terry A. Berger Publisher: Oxford University Press Year: 1995

Thin-layer Chromatography Author: James M. Bobbitt Publisher: Reinhold Publishing Co. Year: 1963

Index

www.ingramcontent.com/pod-product-compliance
Lightning Source LLC
LaVergne TN
LVHW071244210926
848486LV00008B/345